AF782817

Art in Orbit

BIOTECHNE: INTERTHINKING ART, SCIENCE AND DESIGN

Series Editors

Charissa N. Terranova (University of Texas at Dallas, USA)

Meredith Tromble (San Francisco Art Institute, USA)

Biotechne: Interthinking Art, Science and Design publishes books about the history, theory and practice of art and design as they comingle with the natural sciences. The series title reclaims the Greek meanings of the roots bios, conveying life, the living, or citizen-life, and techne, conveying art, skill, or craft. Biotechne thus names the folding of "art" and "science" into complex and hybrid practices that transcend a human-centered "engineering" worldview. "Interthinking," a neologism invented by art and science visionary György Kepes, describes knowledge informed by ecological, systemic, and cybernetic connections, defining the active engagement among fields central to the Biotechne series. This engagement is the source of the cultural creativity and resourcefulness necessary to thrive in the rapidly changing world conditions of the Anthropocene. Biotechne welcomes proposals treating art and design subjects from any time period, antiquity to the present, which speak directly to these contemporary concerns. We seek inventive, cross-pollinating works about the arts and their engagement with sciences from astrobiology to zoology, wherever that engagement occurs, in art or design studios, scientific laboratories, natural habitats, the museum and gallery worlds, performance spaces, medical practices, and the political realm. By identifying significant intersections of art, humanities, and science, and tracking rigorous paths through the cross-disciplinary information jungle, Biotechne serves audiences of both experts and lay readers while substantiating the role of aesthetic insight within the natural sciences.

Advisory Board

James P. Crutchfield, Distinguished Professor of Physics, University of California at Davis, and President, Art and Science Laboratory, USA

Deboleena Roy, Professor of Neuroscience and Behavioral Biology (NBB), Emory University, USA

Sha Xin Wei, Director of the School of Arts, Media, and Engineering, Arizona State University in Phoenix, USA

Titles in the Series:

D'Arcy Wentworth Thompson's Generative Influences in Art, Design, and Architecture: From Forces to Forms, edited by Ellen K. Levy and Charissa N. Terranova (2021)

Sea Currents in Nineteenth-Century Art, Science and Culture: Commodifying the Ocean World, edited by Molly Duggins and Kathleen Davidson (2023)

Plants by Numbers: Art, Computation, and Queer Feminist Technoscience, edited by Jane Prophet and Helen V. Pritchard (2023)

Space Feminisms: People, Planets, Power, edited by Marie-Pier Boucher, Claire Webb, Annick Bureaud, and Nahum (2024)

Art and Biotechnology: Viral Culture from CRISPR to COVID, edited by Claire Nettleton and Louise Mackenzie (2024)

Viral Behaviors: Viruses and Viral Phenomena across Science, Technology, and the Arts, Roberta Buiani (2024)

Organic Modernism: From the British Bauhaus to Cybernetics, Charissa N. Terranova (2024)

Art in Orbit

Art Objects and Spaceflight

Barbara Brownie

BLOOMSBURY VISUAL ARTS
LONDON • NEW YORK • OXFORD • NEW DELHI • SYDNEY

BLOOMSBURY VISUAL ARTS
Bloomsbury Publishing Plc, 50 Bedford Square, London, WC1B 3DP, UK
Bloomsbury Publishing Inc, 1385 Broadway, New York, NY 10018, USA
Bloomsbury Publishing Ireland, 29 Earlsfort Terrace, Dublin 2, D02 AY28, Ireland

BLOOMSBURY, BLOOMSBURY VISUAL ARTS and the Diana logo are trademarks of Bloomsbury Publishing Plc

First published in Great Britain 2025

Copyright © Barbara Brownie, 2025

Barbara Brownie has asserted her right under the Copyright, Designs and Patents Act, 1988, to be identified as Author of this work.

Cover design: Elena Durey
Cover image: Rong Shi, *Symphony of the Cosmos: Explorations on Planetary Molecular Music*, 2023. © Rong Shi, 2023

All rights reserved. No part of this publication may be: i) reproduced or transmitted in any form, electronic or mechanical, including photocopying, recording or by means of any information storage or retrieval system without prior permission in writing from the publishers; or ii) used or reproduced in any way for the training, development or operation of artificial intelligence (AI) technologies, including generative AI technologies. The rights holders expressly reserve this publication from the text and data mining exception as per Article 4(3) of the Digital Single Market Directive (EU) 2019/790.

Bloomsbury Publishing Plc does not have any control over, or responsibility for, any third-party websites referred to or in this book. All internet addresses given in this book were correct at the time of going to press. The author and publisher regret any inconvenience caused if addresses have changed or sites have ceased to exist, but can accept no responsibility for any such changes.

A catalogue record for this book is available from the British Library.

Library of Congress Cataloging-in-Publication Data
Names: Brownie, Barbara, author.
Title: Art in orbit : art objects and spaceflight / Barbara Brownie.
Description: London ; New York : Bloomsbury Visual Arts, 2025. |
Series: Biotechne: interthinking art, science and design | Includes bibliographical references and index. | Summary: "This book explores the contexts, questions, challenges and opportunities for creative exploration of form, materials, and the body, in space. Presenting 9 original case studies from artwork shaped by the unique physical and psychological conditions of space, and informed by exclusive interviews with artists working in the field, it highlights collaborations between artists, engineers, and theorists that have recontextualized the perception and use of weighted materials and subject positions in art practice"– Provided by publisher.
Identifiers: LCCN 2024059570 | ISBN 9781350451735 (hardback) | ISBN 9781350451773 (paperback) | ISBN 9781350451759 (epub) | ISBN 9781350451742 (pdf)
Subjects: LCSH: Art–Philosophy. | Space environment.
Classification: LCC N66 .B77 2025 | DDC 701/.05–dc23/eng/20250206
LC record available at https://lccn.loc.gov/2024059570

ISBN: HB: 978-1-3504-5173-5
ePDF: 978-1-3504-5174-2
eBook: 978-1-3504-5175-9

Typeset by Integra Software Services Pvt. Ltd.
Printed and bound in Great Britain

For product safety related questions contact productsafety@bloomsbury.com.

To find out more about our authors and books visit www.bloomsbury.com and sign up for our newsletters.

CONTENTS

ILLUSTRATIONS

ABBREVIATIONS

AI	Artificial Intelligence
AR	Augmented reality
CNES	*Centre National D'Etudes Spatiales* (French Space Agency)
CSA	Canadian Space Agency
ESA	European Space Agency
ESTEC	European Space Research and Technology Centre
ETTAS	European Space Agency Topical Team Arts and Sciences
GEO	geostationary orbit
IAF	International Astronautical Federation
ISS	International Space Station
JAXA	Japan Aerospace Exploration Agency
JPL	Jet Propulsion Laboratory
KSEVT	Cultural Centre of European Space Technologies (Kulturno središče evropskih vesoljskih tehnologij)
LEO	Low Earth orbit
NAE	Numerical Aerodynamic Simulation
NASA	National Aeronautics and Space Administration
NASDA	National Space Development Agency of Japan
ODF	Operations Data File
SEI	Space Exploration Initiative
UKSA	UK Space Agency
UN	United Nations

Introduction: Art Objects in Orbit

"Humans are shaped by our relationship to Earth," notes artist Alexander Mir (2017: 152), reflecting on his discussions with would-be collaborators in the space industry. This is true physically and culturally: our bodies are shaped by Earth gravity and the conditions in which we grow and procreate; earthly metaphors pepper our languages; borders are dictated by geographical features such as coasts and mountain ranges; cultures are shaped by the climates in which they have developed and the resources available at those sites; creative practice has taken the Earth as inspiration and as a source of materials. Despite the inextricable links between human culture and our home planet, humankind also seeks inspiration from the cosmos. We look outwards and upwards, imagining the vast emptiness of space as a potential site for human activity that is unconstrained by the environments that have shaped Earth-bound practice.

Art has a history of transgressing borders. If space is the final frontier, then the border between Earth and space might be one of the most significant that art has ever transgressed. The expanded spaceflight industry encompasses manned and unmanned activities that collectively contribute to the establishment of space as an environment that can host all aspects of human social and cultural life, including the arts. Through space exploration, new sites have been created for arts practice, and new opportunities to reflect upon what it means to be human; the relationships between ourselves and our planet; and the ways in which art practice has historically been shaped by terrestrial environments. Challenging notions of what it means to be human, space art might be the ultimate cultural resistance.

During the first Space Race and throughout the first few decades of manned space exploration, spaceflight activities were dominated by data collection and the need to test our technical capacity. Increasingly, the themes of discourse surrounding spaceflight have begun to shift from space exploration to "space culturalization" (Turšič 2014). As we begin to colonize space, we will inevitably take with us aspects of human culture. In the era of space tourism and colonization, it is necessary to consider all aspects of the human physical, emotional, social, intellectual, and cultural experience of space. Sectors contributing to the expansion of spaceflight infrastructure have called for increased engagement with the arts. The International Academy of Astronautics (2009: 12) has argued that the role of the artist is essential in the future of manned spaceflight. It warns against the dangers of presuming that the artist's role could be limited to "merely decorative, aesthetic, or provocative 'add-ons'," advocating for the inclusion of artists as essential participants in future spaceflight initiatives. It is in this context that this book explores the role that artists may play in space exploration, and the considerations that may shape their work.

Writing about art in space, it is necessary to unpick the term "space." It is likely that many readers of this book will have found, when searching for this book, texts about *spaces* or *spatiality*—architectural spaces, geographical spaces, social spaces, gaps, and liminal in-betweens. Typically, these spaces are contained, internal spaces within walls, frames, and other defined boundaries (Brandt 2013). It is for this reason that the main title of this book does not contain the word "space." "Space" is an ambiguous term with many meanings, but not quite a homonym, because these meanings are all extensions of the same idea; that the whole universe can be divided into objects and spaces—emptiness and the things that are situated in it. Outer space, and its vast emptiness, exists on a continuum with the more contained architectural spaces that man has made on Earth. Humans carry with us, into outer space, this understanding of space as absence, in opposition to presence, waiting to be filled. Space is a site or location, but it is also an idea. We imagine space as a void, and are fascinated by how this imagined emptiness sits around and within art, as well as how art represents it. In reality, however, space is not empty. Quantum theory tells us that space is full of "dark matter"—particles that "continually form and then disappear" (NASA 2024). Significantly for how artists interact with their materials, space is also full of forces. The force of gravity distinguishes our experience of space on Earth from how it is experienced during spaceflight, where objects feel weightless (see Chapter 3). Space, therefore, is significant for artists both philosophically and physically.

Annick Bureaud (2004) observes the difficulties encountered when trying to define and locate space, and how humans have historically located or conceived of the relationship between Earth and the heavens. She observes that English lacks an equivalent term to the French "*espace*," that describes "the world beyond the world." Conceptually, this notion of space as everything beyond the world is helpful; however, we must also acknowledge that our world is itself located *in* space (White 2022). To say that something is *in space* therefore, does not necessarily distinguish it from something that is *on Earth*, or, indeed, on the surface of another planet. *Orbital space* can be more narrowly defined as being beyond Earth. Orbit is not a place, but rather, a path traced by objects that are located close enough to a planetary body to be pulled toward it, and yet far enough away, or moving so quickly, that they do not fall to the ground. As such, even in orbit, the relationship to Earth is inescapable.

Most human space activity takes place in or below Earth orbit, always close enough for the Earth to remain the dominant feature in view. The title of this book, *Art in Orbit*, refers to work that is beyond Earth, and yet always within Earth's shadow. This is representative of where we are, at present, in humankind's journey toward becoming an interplanetary species. We are only just beginning to engage with space, to develop human space culture, still understanding the experience of space through contrast with terrestrial experiences. *Art in Orbit* begins to establish what art might be possible in space, as artists challenge their terrestrial ways of thinking and working, in the immediate vicinity of Earth, where such works will inevitably be understood through contrast to terrestrial art.

For most of the artists featured in this book, the conditions, rather than the location of space, define the site of their practice. Art that engages directly with the physical conditions of space can be more meaningful to artists than the altitude at which the work is produced or deployed. This understanding of spaceflight activity is echoed by NASA (2021: 3), who collectively refer to high-altitude balloons, parabolic aircraft, and rockets as "suborbital flights" regardless of their altitude. When this book describes the environment of space, it refers to the conditions of spaceflight, principally when passengers become weightless or are presented with the "overview" (see Chapter 1). These conditions are encountered at different altitudes, depending on the nature of the spaceflight, including onboard reduced gravity flights that fly well below the Kármán line that the Fédération Aéronautique Internationale defines as the boundary between Earth and space. Orbit is just one site at which these conditions exist, and is helpfully representative as it is, at present, the site of most human space activity. As the spaceflight industry evolves, more will venture beyond Earth orbit, and a range of different conditions will be encountered. When activity beyond Earth orbit becomes more common, future theorists and practitioners may need to revisit the definitions and considerations outlined in this book, in light of new discoveries and experiences.

As this book will show, space art is most significant when it directly responds to the conditions of space, as in, for example, the book's first case study, *Impossible Object* (2022), in which artists Yasmine Meroz and Liat Segal have approached the environment of space as a medium. I have chosen to examine *objects*, as opposed to the arts more broadly, as these are more substantially affected by the environment of space than other artforms. The term "objects" is used to describe sculptures, as well as materials that become sculptural in microgravity, and other three-dimensional artefacts, including artefacts and materials that may become three-dimensional or adopt the form of an object in space. Some artforms, such as images, are not altered by the environment of space, and so approaching to the production of images is less altered by the potential for deployment in space than practices relating to three-dimensional objects. That said, this book does not discount the process of image production. Examples featured in this book include images where their production is affected by microgravity, and discussion of the orientation of images (see Chapter 4) is applicable to flat images. The production of images in space is affected by microgravity when the qualities and behaviors of materials used in their production are affected by gravitational conditions (see Chapter 6). In such instances, however, it is not the image itself that is most significantly affected, but, rather, the materials that, as a result of microgravity, become sculptural and *object*-like.

The *objects* in this book include *bodies*. This includes human bodies as well as non-human, inanimate bodies. This mirrors the practice in the space exploration community of using the terms *bodies* and *objects* interchangeably in some contexts. The term, *object*, has significance in space exploration more broadly. Astronomy is concerned with "celestial objects," or "celestial bodies" (Jardetzky 2005 [1958]; Beletsky 2001). The largest of these are stars and planets that are shaped and propelled by gravity. There is a kind of passivity in the way that celestial bodies or objects move, not being propelled by their own means but by the force of gravity, for example, as one or more stars pulls a planet into its orbit. In our own single-star solar system, planets move in simple elliptical orbits, but where there are two or more stars in a system, where orbit is a dance with many partners, and a planet's orbital path may be complex (Beletsky 2001: 3). In space, objects including artificial artefacts and human bodies behave in the same way, being shaped and/or propelled by gravity in ways that are unfamiliar to terrestrial beings who take groundedness for granted. Astronauts and performers who have experienced weightlessness describe the loss of control that comes with floating in the void, in which body posture becomes "neutral" and "you let yourself be taken by the environment" (*sic*) (Dubois 2003 [1999], see Chapter 6).

The use of the term *object* in reference to human as well as inanimate bodies does not wholly reject anthropocentrism. Arguments for transporting art into space are dependent on an anthropocentric view of the importance of human culture, and the value of art objects as expressions of what it means to be human. Space travel prompts questions about human identity, and this extends to questions about the nature of life; the distinction between living and not-living; animate and inanimate objects. Yet, the study of space is object-oriented; human spaceflight highlights the interrelatedness of objects outside the sphere of human influence, including the forces that objects exert on one another, and including the experience of the body as object-like when it loses control. In short, space is not shaped by human hands to the degree that our terrestrial environment is, and humankind has not yet adapted to spaceflight sufficiently to have as much agency in the environment of space as it does on Earth.

In pursuit of an understanding of space art, this book will begin, in Chapter 1, by describing the current state of space exploration and how art fits into the bigger picture of the "era of space experience" (Spencer 2022). Describing how space art has emerged alongside space exploration, it surveys the evolution of space art from *astroart*—terrestrial art that is inspired by space art—to *spaceworks* that are produced or deployed in the environment of space. It introduces various definitions of space art and related practices that engage either directly or indirectly with space, and aligns these with the space sector activities that make such art practices possible, including the spaceflights and orbital space stations that enable the production of art in orbit.

Chapter 2 explores the cultural contexts that shape, enable, or sometimes limit space art, primarily focusing on the art/sci context. The chapter observes the extent to which the space sector has been dominated by science and technology, causing space art to be framed as an outreach activity, produced by hobbyists as opposed to professionals and specialists. The chapter records how artists producing space art have had to reframe their art projects as science experiments in order to gain approval for launch, but also, more positively, how new initiatives are beginning to change the perception of space art and assert its value. This chapter considers knowledge-exchange between the arts and space sectors, and the different ways that knowledge is acquired by scientific and arts communities, showing that the space sector can learn much not only from spaceworks themselves, but also in the methods of enquiry that are deployed by space artists.

Chapter 3 begins to explore the experiences and environments that shape space art, starting with weightlessness. Weightlessness is the most otherworldly aspect of the spaceflight experience, significantly affecting objects and the ways that they are encountered. This chapter establishes the view that weightlessness necessitates radical new approaches to creative practices, and prompts "postgravity" thinking. Postgravity art recognizes the extent to which humankind and our creations have been shaped by Earth gravity conditions, and invites us to revisit many of the approaches that underpin terrestrial art. This chapter considers the location of spaceworks' audiences, since the audience may not experience the same conditions of weightlessness as the art itself, and explores ways that the experience of weightlessness may be represented for those Earth-bound audiences.

Chapter 4 addresses the topic of orientation. Traveling away from Earth, spaceflight participants enter an environment of subjective orientation, in which there is no up or down. This chapter records a brief history of how spaceflight has prompted a shift from the dominance of upright orientation, first to aerial views accessed via aircraft and satellites, and then non-oriented views experienced onboard space stations. Examples such as Arthur Wood's *Cosmic Dancer* will be introduced to illustrate how non-oriented approaches may be manifested in spaceworks.

In Chapter 5, this book considers the phenomenology of space and the effect of weightlessness on the human body. This chapter draws on the experiences of performers and astronauts to consider how the human body becomes object-like in space, and the relationship between body and ground that is taken for granted on Earth's surface. The chapter introduces examples of performance art, in particular by choreographer Kitsou Dubois, as well as art that becomes performance as a result of its taking place in microgravity. It considers the experience of being suspended in a void, as well as physical relationships between bodies and objects that are not held together by gravity.

Chapter 6 examines the crafts and materials that artists may employ in the production of spaceworks, focusing on how the environment of space changes the behavior of familiar materials. It considers the craft practices that underpin art, and how those practices are affected by microgravity. It then establishes some of the different behaviors of materials including artists' media in the environment of space, illustrating how liquids can become sculptural when weightless, and how astronauts' scientific experiments have closely informed those of artists. The chapter goes on to argue that, since tools and materials behave so differently in space, there is scope to develop new tools for spacework production.

While these main chapters present the contexts and concerns that shape art in the environment of space, between them can be found case studies presenting examples of real-life engagements with that environment, illustrating the current state of the field and the concerns that have informed recent *spaceworks*. These case studies are informed by interviews with artists who have deployed work in the environment of space—onboard the International Space Station, in open space, and on reduced gravity aircraft—in each case responding to a particular set of circumstances but collectively sharing an ambition for the arts to be more actively engaged with phenomena encountered during spaceflight. All of these case studies exist within the contexts that are explored within the book's main chapters, and hence are distributed between the chapters, appearing after the chapter deemed most relevant.

The artists interviewed for these case studies range from some who have had no formal engagement with space science to those who consider themselves scientists as much as artists. Some have made it their ambition to produce work for space, while others found themselves unexpectedly with an opportunity to do so. There are few artists who specialize in producing spaceworks. Eduardo Kac, interviewed for Case Study 6, is one of few artists to have deployed multiple works in space, and perhaps the only artist to date who may be considered an experienced specialist in this field. Others have found themselves limited to just one or two opportunities, meaning that their work is necessarily experimental in nature.

Trevor Paglen is the only artist in these case studies to have produced work specifically for open space. *Orbital Reflector* (Case Study 1) was launched, but never fully deployed, illustrating the extent to which it is still difficult to produce art satellites. Others, including works by Yasmine Meroz and Liat Segal (*Impossible Object*, Case Study 2); Max Baraitser Smith (*Lunar Bureau of Weights and Measures*, Case Study 3); Lisa Pettibone (*Verdant*, Case Study 4); Eyal Gever (*#Laugh*, Case Study 5); and Eduardo Kac (*Télescope Intérieur*, Case Study 6), were deployed in orbit onboard the International Space Station. Due to the youth of this field, not all of the case studies and examples contained within this book are sited in orbit. For the final three—Jeanne Morel (Case Study 7); Alexis Hope (*Space/Craft*, Case Study 8); and Sana Sharma (*Fluid Expressions*, Case Study 9)—reduced gravity aircraft are used as a substitute for, or in preparation for orbital spaceflight. This is necessary at present because access to space is still so restricted that much of the art that is intended for space may only be deployed in environments that simulate space conditions. These examples are no less valuable than those that are deployed in orbit, as they all contribute to establishing the foundations of this emerging field. In contrast to artists whose work has been deployed in open space onboard the International Space Station, where work must be deployed on the artist's behalf by an astronaut, those artists who have worked onboard parabolic flights have had the opportunity to deploy the work themselves and learn from this hands-on experience. Due to the accessibility of parabolic flights relative to spaceflights, both Morel and Sharma have had opportunities to refine their practice across several flights.

To locate these case studies, and the other works referenced in this book, in their contexts and in relation to each other, this book concludes with a timeline of all cited works. The data presented in this timeline broadly paint a picture of the extent of the field of space art at the time of writing and the pace of its development. Thus, the cited examples are representative of the emerging field of space art. It remains to be seen how this field may expand and diversify over the coming decades.

1

Art in the Commercial Space Age

From astroart to space art

Space is now a site for creative production, and for the deployment of space art. Roger F. Malina (1991: 147) defines "space art" as "contemporary art which relies for its implementation on participation in space activity." As will be examined later in this chapter, Malina's definition is both helpful and unhelpful in equal measure. Malina acknowledges that there is potential for a field of arts practice that is linked to space exploration, but, as this definition encompasses such a broad range of practices, it is difficult to use Malina's definition alone as the basis for identifying the concerns that might shape this emerging field. Since the publication of Malina's text, the field has expanded sufficiently to demonstrate the need for more granular definitions of the kinds of practice that space art may encompass. Future creative engagements with space may take many forms, and practitioners in a range of art and design disciplines associate their practice with space and the cosmos. These inter-related practices have become too diverse to be the subject of a single exploration such as the one presented in the book. It is necessary to begin to distinguish these practices from one another, and to identify the concerns that are unique to each field. The possibilities for space art, and the practices that it may encompass, are multiplied in the commercial space age, when artists have opportunities to engage with space in a variety of ways, both directly and indirectly.

This book aims to establish distance between contemporary *space art*—made possible by increased access to the environment of space—and *astroart*, which is a step removed from spaceflight. "Astronomical Art," more often abbreviated to "astroart," is, broadly speaking, "creative [practice] inspired by astronomy," including science-fiction art (De Paulis 2014; Hughes 1978: 326; Heck 2007). Astroart pre-dates manned spaceflight, and is typically representative in nature, tending to feature depiction of celestial bodies and fictional or fictionalized space exploration activities. Astroart thrives today in part due to governmental spaceflight organizations' public outreach initiatives, such as National Aeronautics and Space Administration's (NASA's) *Hubble Heritage Project* (1998–2016), which sought to produce "aesthetically impactful and scientifically accurate color images of our universe ... fostering a deep and lasting appreciation of astronomy within the public." Such initiatives are indicative of a view that is pervasive in the space exploration community, that it is the principal responsibility of the astro-artist to assist space exploration organizations to inform a public audience about their discoveries, and to incite public interest. As Chapter 2 will show, a tendency for the spaceflight industry to equate space art with representative art, including astroart, reinforces unhelpful assumptions about the ways that the creative and spaceflight industries can collaborate.

Throughout the first Space Age (a period that broadly encompasses the Space Race and the early days of space exploration, from the late 1950s to the 1970s), the arts demonstrated ambition to become involved in space exploration, but for much of the history of spaceflight artists have been excluded from direct participation. Much of the art produced during, or associated with, the first Space Age, can be classified as astroart. Space Age artists looked toward space for inspiration, but without opportunities to become directly involved in spaceflight, they were only able to engage with space symbolically and indirectly. Art of the Space Age is associated with a Space Age aesthetic, tending to represent space and spaceflight either through direct depiction or through rocket-like or celestial forms and metallic colors, without consideration for the ways in which the environment of space might directly shape art production (Carroll 1982; Chang 2022). An early exception is Paul Van Hoeydonck's tin statuette, *Fallen Astronaut* (1971), that was placed by Apollo astronauts on the surface of the Moon as a memorial to deceased astronauts and cosmonauts (Pocock 2012: 371). More recently, opportunities have arisen for direct engagement with spaceflight through collaboration with spaceflight providers, resulting from the expansion of the industry and initiatives to democratize space. The current era of commercial spaceflight offers new opportunities for artists to go beyond representation and toward direct engagement. Through this direct involvement in spaceflight, it is possible to produce work that would not be possible on Earth.

The works explored in this book engage directly with the conditions of space. These are works that could not behave or be experienced in the same way in the terrestrial environment. This includes, for example, work that is produced within microgravity conditions using materials that behave differently when weightless, or works whose situation or presentation is made possible by microgravity. Jason Dunn, co-founder of Made in Space, goes so far as to propose that, hypothetically, art in space might include "new types of art that ... we can't even bring back to Earth because [they] wouldn't even survive in [Earth] gravity" (Gever 2016). This definition excludes work such as astroart, that represents or observes space from

Earth, or that makes use of data that a third-party has collected from space. In order to directly engage with the conditions of space, it is necessary either for artists to travel to space (or an environment that simulates the conditions of space, such as a reduced-gravity flight), or for their work to be deployed in those environments. Opportunities for the latter have been provided by initiatives such as Arts Catalyst's *Microgravity Interdisciplinary Research* (MIR) initiative (2000–4); MIT Media Lab's *Space Exploration Initiative* (2017–present); and the Israeli *Rakia Art Mission* (2022), which have supported projects ranging from music and dance performances to craft experiments. Meanwhile, other artists have independently realized artworks to be launched into space, deployed either in open space (such as Peter Beck's *Humanity Star*, 2018), onboard spacecraft (such as Arthur Woods' *Cosmic Dancer*, 1993), or on the surface of the Moon or another planet (such as Damien Hirst's *Spot Painting* for the Beagle 2 Mars lander, 2003, or Jeff Koons' *Moon Phases*, 2024).

Due to the youth of this field, many of the works referenced in this book have been proposed, but not fully realized. Artists such as Albert Notarbartolo (1975) proposed deployment of "spaceworks" long before spaceflight organizations had capacity to accommodate such activities. Notarbartolo (1975: 140) proposed sculptural spaceworks that could be placed in geostationary orbit, including *Beacon*, "a hemispherical element, the interior surface of which reflects sunlight toward the Earth through a prismatic element ... made of transparent materials in different colors"; and *Star Cloud*, an array of transparent spheres, each housing a rotating reflective disk. Notarbartolo imagines *Star Cloud* as a potentially infinitely expandable work, with the possibility that future astronauts could contribute additional spheres of varying sizes. Notarbartolo recognized that these proposals were speculative, but imagined that he might live long enough to be given the opportunity to travel into space and create a spacework during his journey. In 1986, the Eiffel Tower Competition challenged artists to design an object that, when launched into orbit, could be visible to the naked eye on Earth (Coles and Jefferson 1989: 201). The works submitted for this competition helped to establish the potential for a field of practice in space art, showcasing what might be possible, while also highlighting potential challenges. Among the selected entries was Pierre Comte's (1987: 18) *ARSAT*, a large square, mirrored sail that would be unfurled in orbit. Group Spirale's competition entry was *L'Anneau de Lumiére*, a 24 km diameter ring of reflecting balloons that would have formed a ring in the night sky larger than the Moon (Raitt 2007: 50). Two common themes emerged in these and other competition entries: reflective surfaces, that would rely on reflected sunlight to make the work visible from Earth; and the ability to unfurl, paving the way for many future works of space art that have been folded or compressed for transport and then unfolded or deployed automatically or by astronauts once in space. Plans to deploy the winning entries of the Eiffel Tower Competition were cancelled following opposition from astronomers, among others, to the presence of objects in the night sky. Arguments for the preservation of the sky continue to affect human space activities today, including proposals for satellites with artistic or scientific purpose (Shepherd 2023).

The barriers to launching art into space have continued to impact the deployment and reception of space art in recent decades. In 2018, Trevor Paglen's Orbital Reflector was launched successfully into space onboard a SpaceX Falcon 9 rocket, but failed to deploy fully (see Case Study 1). The work was a 30-meter-long, reflective "elongated diamond" that was to be inflated from a miniature satellite, and would have been, once deployed, visible at dusk from any location on Earth (Griffin 2018). In the same year, New Zealand-based spaceflight company Rocket Lab successfully launched Peter Beck's reflective geodesic sphere, *Humanity Star,* into low-Earth orbit (see Figure 1.1). When viewed from Earth, the orb resembled a meteor, and was intended to provoke the feeling of wonder that can be experienced when viewing a shooting star. Arguments against pollution to the night sky were countered by plans to make these projects temporary, orbiting the Earth for only a few weeks before falling toward Earth and disintegrating upon re-entry. Nonetheless, opponents objected to *Orbital Reflector* and *Humanity Star* on the grounds of what some viewed as a lack of purpose—"going into orbit for purely aesthetic and philosophical purposes"—a view that is only very slowly being overcome as the spaceflight industry begins to recognize the value of non-scientific space ventures (Griffin 2018; Guarino 2018).

Space stations have proven to be less controversial sites for the deployment of space art that is much smaller in scale. Arthur Woods' *Cosmic Dancer* was deployed on his behalf by cosmonauts onboard the Mir space station in 1993. Woods' geometric, painted aluminum sculpture took advantage of the microgravity conditions onboard Mir by having no defined resting point. In video documentation, cosmonauts can be seen interacting with the work, nudging its corners to adjust its angle and speed of rotation as it revolved through the interior of the space station. While weightless, Woods' sculpture can be seen "in a state of perpetual change" in that its orientation, and hence composition, is continuously altered as it "playfully dances" through the astronauts' habitat (Woods 2021a). A number of artworks have been deployed onboard the International Space Station (ISS). In 2022, the ISS hosted the *Moon Gallery*, a collection of sixty-four miniature artworks (https://moongallery.eu/). The collection includes two-dimensional and three-dimensional works, including some that behave differently in the microgravity conditions onboard the ISS than they would on Earth. Lisa Pettibone's *Verdant* (2022) is one such work, a sculpture of glass beads, tethered to one another with neon yellow thread in a configuration that will alter while the work is weightless (see Case Study 4). A different kind of engagement with microgravity can be observed in Liat Segal and Yasmine Meroz's *Impossible Object* (2022, see Case Study 2), a sculpture that was

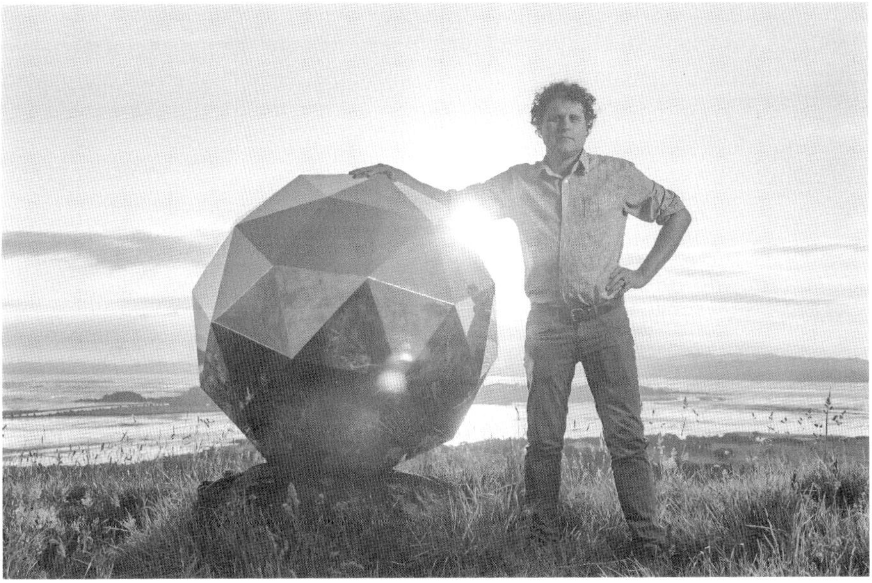

FIGURE 1.1 Peter Beck, *Humanity Star*, carbon fiber geodesic sphere with reflective panels, 3 ft × 3 ft, 2017. The passive satellite was launched into orbit by Rocket Lab in 2018. © Rocket Lab

deployed on the ISS as part of the Israeli Rakia Art Mission. *Impossible Object* is made partly of water, and these parts have form only when weightless. Segal and Meroz (2022) identify microgravity as their medium, and chose to title the work "impossible" because it "cannot exist on Earth." Such works, that cannot be considered complete until they are deployed in space, illustrate that the challenges and opportunities of art practice in the weightless environment are significant in differentiating the concerns of space arts from terrestrial arts, and, as will be explored throughout this book, introduce a range of considerations that are unique to space art.

Humankind in the era of commercial spaceflight

In the first few decades of the twenty-first century, we have entered a new, commercial space age, in which the variety of activities taking place in space has expanded significantly, and the number of spaceflight providers has expanded to include commercial ventures. The commercial human spaceflight industry is currently dominated by three providers, all of whom have successfully transported commercial and civilian passengers into space: Blue Origin, Virgin Galactic, and SpaceX (Leslie 2022). Humans have established a permanent presence in space, with crew onboard the ISS, and, since 2022, the Chinese Tiangong space station. The activities that take place onboard these space stations are predominantly scientific research, but the ISS has also played host to those who seek to use their time in space for creative pursuits, as well as commercial passengers who have traveled to space with the sole aim of using the ISS as the location for the creative production, most notably actress Yulia Peresild and producer Klim Shipenko who spent twelve days onboard to film a science-fiction film, *Вызов* (*Challenge*) (Pearlman 2021). The ISS has also played host to art that has been designed on Earth to be deployed in space, including sculptures by Arthur Woods and Eduardo Kac, both of whom are explored in this book.

David Ashford (2013: 137) lists the establishment of "orbital art studios" as one way that space may further develop as a culture and tourism destination, supported by an infrastructure of accommodation and accessible transport to and from site. Work toward this infrastructure is underway. The field of space architecture is well-established, and members of the space architecture community advocate for further involvement from artists in future spaceflight activity including the development of new orbital habitats (International Academy of Astronautics 2009: 12). The late 2020s and early 2030s should see a significant increase in the number of orbiting space stations, and hence an increase in the variety and availability of orbital sites that may host creative activities. Rather than directly replace the ISS with a single station after it has been decommissioned, NASA has invited other organizations to develop a range of "private outposts" in low Earth orbit, freeing NASA to redirect its resources to exploration of the Moon and Mars (Diner 2022; NASA 2023a). Axiom Space Station is under construction at the time of writing, and is due for launch in 2026, beginning with modules that will be attached to the existing ISS (https://www.axiomspace.com/). A year later it will be joined by Starlab, due to be launched onboard a SpaceX Starship in 2027 (Berger 2024). Within a decade, these may be joined by a range of other private stations including Bigelow Alpha Station and Orbital Reef, and government stations such as India's Bharatiya Anteriksha Station (Pillai 2024). These stations will contribute to establishing low Earth orbit as a site for wide-ranging government and commercial activities across an ever-increasing range of sectors, including, potentially, the arts.

Jan Wöerner, former Director General of the European Space Agency (ESA), identifies this new era in space exploration as "space 4.0." He traces the journey from "early astronomy" (space 1.0) to the Space Race (space 2.0) to "the opening of different applications such as Earth observation and navigation" (space 3.0), and today, in space 4.0, an era of "commercialization and participation" (Mir 2017: 61). In this era of commercial spaceflight, space is a site for a wide variety of governmental and commercial, manned and unmanned activities. Space 4.0 has introduced what space designer John Spencer (2022) describes as an "era of space experience," when an increasing number of spaceflight participants are choosing to travel to space solely for the experience. Research predicting the expansion of the commercial spaceflight industry identified two experiences that motivate a design for space tourism: the experience of viewing the Earth from space, and the experience of weightlessness (Spennemann 2008: 163; Cohen 2017: 27). The nature of the spaceflight experience differs between providers. Sub-orbital flights, such as those offered by Virgin Galactic, grant passengers a short period in which to view the Earth's horizon against the backdrop of space, alongside a few minutes of weightlessness. The flight is short and intense, inviting passengers to reflect on their experience in hindsight after their return to Earth. By contrast, World View balloons will offer a more leisurely six- to eight-hour stratospheric journey that grants passengers a much longer period to reflect on their experience of viewing the curvature of the Earth, but without the experience of weightlessness. Weightlessness can be experienced without the view of the Earth onboard a reduced-gravity flight such as those offered by Zero-G and Novespace, which enter freefall in order to offer several short periods of weightlessness each lasting approximately thirty seconds.[1] More costly and therefore more exclusive experiences are offered by SpaceX, whose orbital flights grant passengers a longer experience of both weightlessness and the view of the Earth.

The next significant development in space experience, predicts Spencer (2022), will be lunar fly-bys. One such planned activity may be the dearMoon mission, a SpaceX flight that will, if it goes ahead, transport eight civilians around the Earth and then the Moon so that they may observe and respond to their experience of Earth and lunar orbit (https://dearmoon.earth/). The mission is funded by Japanese entrepreneur Yusaku Maezawa, who has consciously selected crew that identify as "artists," in the hope that the opportunity would elevate their current creative practice (Maezawa 2021).[2] The selected crew includes two photographers, a YouTuber, a DJ, a filmmaker, a musician, an actor, and a choreographer. Maezawa's (2022) introductions to his selected crew seem to suggest that he has a particular interest in the work that these artists will produce during the flight, as a collective, mixed-media record of the mission, as well as the ways in which the experience will inform their future practice (see Chapter 2). Maezawa has

---

[1]There are differences of opinion as to whether all of these activities constitute spaceflight. Virgin Galactic flights are a particular focus of this debate. While NASA is of the view that Virgin Galactic flights do constitute spaceflight as they operate beyond the mesosphere, others argue that they do not constitute spaceflight as they do not pass the Kármán line that is identified as the start of space by the wider scientific community (Perez 2021). World View balloons and reduced gravity aircraft operate at lower altitudes and do not claim to offer spaceflights but are typically categorized as space tourism as they offer experiences akin to those of true spaceflight—the view of the curvature of the Earth and/or weightlessness (Webber 2019: 165).

[2]In order to assess candidates' eligibility for the flight, it was necessary for Maezawa to define "artist" in his own terms. Acknowledging the ambiguity of the term, Maezawa (2021) concluded that he would invite applications from anyone who engages in creative practice. This will be explored further in Chapter 2.

previously traveled to the International Space Station as a space tourist, and stresses that the dearMoon flight will offer distinctly different experiences, including the opportunity to witness Earthrise as the Starship passes around the Moon.

Space experience, in contrast to scientific and exploratory spaceflight, is distinctly human. Space travel has the potential to prompt profound reflections on what it means to be human. Spaceflight is a sublime experience full of contrast: light and dark, noise and silence, motion and stillness, infinite distance and uncomfortable proximity (Cohen 2017: 26; White 2021: 27). Actor William Shatner, following his short voyage onboard a Blue Origin flight in 2022, expresses how unprepared he had been for these contrasts and their profound consequences. In his autobiography, Shatner (Shatner and Brandon 2022) describes the extreme pressure of two- and three-G gravitational forces experienced during launch, followed by the release of weightlessness, then, two contrasting views visible from the spacecraft's windows: one of life and light, and the "warm, nurturing … Earth below"; and one of "cold, dark, black emptiness" in the "deep, enveloping, all-encompassing" space on the other side of the spacecraft. Interviewed on *Today*, he describes being "overwhelmed with the experience of looking at death and looking at life." Far from the "majestic awe" that he had anticipated before his journey, the experience filled him with "overwhelming sadness" (*Today*, n.d.).

Spaceflight is both awe-inspiring and terrifying—transcendental, and yet visceral. These contrasts arise in part from our reliance on technology that is necessary to enable human spaceflight, that acts as a constant reminder of the physicality of the human body and the effort required to sustain life in a lifeless environment. While space is vast, empty, and silent, spaceflight begins with the roar of rocket engines, astronauts' bodies are compressed into uncomfortably compact spaces, densely populated with tools and equipment. Comparisons can be made to the experience of clinging to a small raft in the middle of the Atlantic Ocean, or sinking to the bottom of the Mariana Trench on a tiny submarine, with the size of the surrounding ocean infinitely multiplied. Astronaut Chris Hadfield describes how the experience of space is mediated through a spacesuit, in which the sound of the fan and ventilation system and of the astronaut's own breathing and pulse act as constant reminders of the human bodily functions and their dependence on life-sustaining technology (Savage 2015). These mechanical and biological sounds are alien in space, intimate in ways that are native to Earth, characterized by their closeness in contrast to the silent expanse of space.

Spaceflight reportedly provokes "a shift in World View" that Frank White (2021: 5) describes as the "Overview Effect," in which space travelers' experience transforms their perspective on what it means to be a human, and an Earthling. White describes how the view of the Earth from a distance prompts critical reflections on the extent to which aspects of Earth habitation are taken for granted, through "a profound understanding of the interconnection of all life, and a renewed sense of responsibility" for the planet. White stresses the significance of the ISS's cupola in providing astronauts with an opportunity to develop a "new frame of mind" as they gaze upon the Earth from afar from this optimally located window (White 2021: 63; 54). Astronaut Jim Irwin suggests that the strength of this feeling is multiplied with distance, in particular where other planets are visible and their inhospitality may be directly contrasted with the hospitability of Earth (White 2021: 51). The effect can be brought on by direct and vicarious experiences of space travel, not only by the sight of the Earth from space but also by verbal and visual records of this experience (White 2021: 78). During the Apollo 8 mission, astronaut William Anders captured a photograph that has since become known as *Earthrise* (1968). This image of Earth from afar prompted a newfound appreciation of Earth as a fragile store of finite resources, and informed the establishment of the first global environmentalist movement (Deese 2009: 70–1). Years later, the Voyager 1 space probe left our solar system and transmitted an image of the Earth as a "a lonely speck in the great enveloping cosmic dark." Seeing this image, Carl Sagan (1994: 7) wrote that "our posturings, our imagined self-importance, the delusion that we have some privileged position in the Universe, are challenged by this point of pale light."

The Overview Effect causes many astronauts to return to Earth with an increased interest in ecology and the preservation of the planet, and for this reason White, among others, advocates for orbital environmental summits and increased opportunities for space travel that might have the effect of strengthening humankind's resolve to preserve the Earth's climate (White 2021: 92; 93; 101). White's views prompt consideration of the role of spaceflight in the context of the climate crisis. The relationship between climate activism and the spaceflight industry is riddled with contradictions. While spaceflight paves the way for numerous proposed solutions to the climate crisis, including, for example, space-based solar power, environmental researchers predict that space tourism poses a significant threat to Earth's climate (Frazer-Nash Consultancy 2021; Ryan et al. 2022). Many spaceflight providers, responding to concerns about the overpopulation of Earth and the climate crisis, identify spaceflight as the first step toward the colonization of other planets, either in a bid to preserve human existence beyond the lifespan of Earth as a habitable planet, or to exploit the resources that other planets have to offer to relieve the pressure on finite resources of Earth (Dunnett et al. 2019: 320).

There is a risk that the future of space exploration may follow the same pattern as historical exploration of the Earth, beginning with exploration and leading to colonization and exploitation. Since John F. Kennedy's 1960 presidency acceptance speech drew parallels between the frontiers of space and the American West, space has been described as a new or final "frontier." Central to this discourse is the image of space as infinite, echoing a "colonial promise of limitless growth" (McCormick 2021). While we are at risk of depleting the finite resources of Earth, space appears to have the potential for "consequence free" exploitation. However, "the frontier metaphor," argues Lisa Messeri (2017), "sets up outer space as a passive landscape with no purpose other than human sustenance." The spaceflight industry is already suffering the consequences of this over-simplification, as orbital space is fast becoming cluttered with debris that has the potential to endanger future space travelers (Klinkrad 2006). While the cosmos might indeed contain all the physical resources that humankind may ever require, we also need to be conscious of the consequences of an exploitation-driven space exploration and the human capacity to pollute the environments that we exploit (Dunnett et al. 2019: 321–2).

The position of art and artists in space cannot be disentangled from the many other human social and cultural concerns that shape our perceptions of space colonization. While access to space is reserved for the privileged few, it is nonetheless still largely perceived as a public space (Gérin 2003: 36).[3] Much space art is arguably, therefore, public art. Malina (1991: 146) imagines that some space art may be visible to global audiences. Art located in orbit may be geostationary, resting above a fixed point on Earth, or, like the ISS, may move across the sky, becoming visible above different nations as they complete their rotations around the globe. Even while governance of low Earth orbit is contested and organizations seek to lay claim to extra-terrestrial territory, "the notion of limitless, ever-changing space" tells us that there is space to spare, and space for everyone (Gérin 2003: 36). The Overview Effect contributes to the impression that space should be free from political and national divisions, as the view from the Earth from space is of "one world"—a shared home for all humankind, with "no dividing lines between people" (White 2021: 54). This ideal is manifested in Xu Bing's 卫星上的湖泊 (*Satellite Lake: Cosmic Reflections* 2021–22, see Figure 1.2), an animation that was uploaded to, and then displayed on, Commsat's Ladybug-1 satellite after it had been decommissioned but remained in orbit.[4] The animation depicts a human figure carrying a sack.

[3]The United Nations' (UN's) 1967 *Treaty on Principles Governing the Activities of States in the Exploration and Use of Outer Space, including the Moon and Other Celestial Bodies* asserts that "exploration and use of outer space should be carried on for the benefit of all peoples" and that "outer space ... is not subject to national appropriation by claim of sovereignty."

[4]Commsat, a Chinese private satellite provider, intended for the Ladybug-1 satellite to serve multiple functions including those described as "entertainment." For this purpose, the satellite is equipped with an LED display and selfie stick, and a high-powered light array that enables the satellite to twinkle morse-code messages for stargazers on Earth. Entertainment uses of the satellite included, for example, the use of the selfie function to shoot images for a poster for the science-fiction film *Wandering Earth* (Commsat 2019).

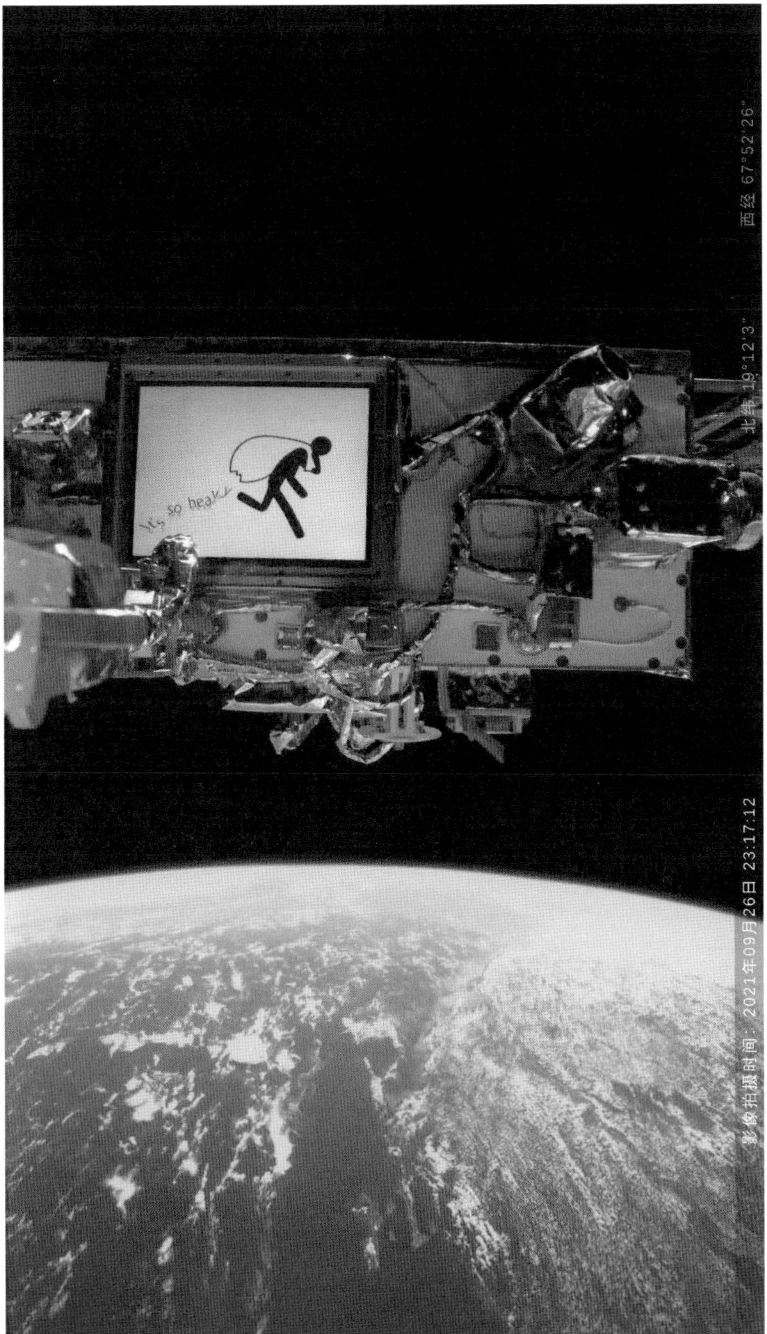

FIGURE 1.2 Xu Bing, *Satellite Lake: Cosmic Reflections*, 2021–(still), 2:08. © Xu Bing Studio

The sack is torn, so that its contents—words—spill out and form a trail behind the figure. Those words appear in a variety of different languages, changing as the satellite passes above corresponding locations on the Earth's surface. Bing's work has long explored the role of written language in cross-cultural exchanges, and in this orbital work he is able to explore this theme in ways that are beyond international, or even extra-national. The work's journey is unobstructed by international borders, highlighting that the relationship between language and place is arbitrary. Traversing the sky, such works demonstrate that space may be a site in which artists can respond to notions of national verses global identity, cultural and political divisions, and the potential to escape these precepts through spaceflight. Such works seek to challenge the overt nationalism that is visible in space exploration, where "the geographies of outer space are inherently linked to terrestrial understandings of [nationhood]" (Dunnett et al. 2019: 330; Lee 2018). This contradiction has implications for artists, their funders, and the ways in which space art is perceived.

The complexities of the human relationship with Earth and space are illustrated in the efforts made in 2016 by the Moderna Galerija in Ljubljana, Slovenia, to collect and exhibit work on the theme of the cosmos for its eighth Triennial of Contemporary Art, entitled *Beyond the Globe*.[5] The exhibition seeks to show that humans are "increasingly involved with the cosmos," and collects works that share the common theme of "contemporary art in the spirit of coming interplanetariness" (Badovinac 2016: 4–5). The exhibition collects a small number of works that explicitly engage with the theme of interplanetariness and exhibits them alongside works that have been reframed in this context, to a greater or lesser degree. Works including Tanja Lazetic's *The Skies Above Me* (2015–16)—photographs of the night sky cropped into circles to represent black holes—and Sašo Sedlaček's *Sky in Ruins* (2016)—a ceiling projection that depicts the collision of satellites and space trash in orbital space—can be located in the tradition of astroart. Other represented artists have sought ways to engage with the ways in which environments beyond Earth alter the physical properties and behaviors of art media, and the experience of human physicality. Most of these explorations are philosophical rather than practical, as in Goran Bertok's *Visitors* (2005), a series of photographs of human corpses burning during cremation that ponders the physical limitations of the human body in the context of "the dictatorship of gravity" (Bertok 2016: 31). Other artists reflect upon their previous work in the context of the curator's interplanetary themes, including

[5]The 8th Triennial of Contemporary Art U3: Beyond the Globe, was curated by Boris Groys and exhibited from June 3 to September 18, 2016, at the Moderna galerija, Ljubljana. Curator Boris Groys (2016: 14) observes that many of the artists represented in the exhibition were inevitably influenced by universalist ideals, informed by a Soviet history of space exploration and ambitions toward uniting humankind in collective human endeavours to become an interplanetary species.

Staš Kleindienst (2016: 81 who draws parallels between "the endless expanded of the cosmos" and "art" that is located in an infinite "universe of cultural exchange. These works are juxtaposed with those that are only loosely and retrospectively related to the theme of interplanetariness, being primarily concerned with contemporary questions about what it means to be human, or post-human, in the era of artificial intelligence (AI) (Keti Chukhrov's *Love Machines*, 2013); surveillance (BridA collective, *Tracked Houston*, 2014); or DNA mapping (Gregor Mobius' *DNA Semantics*, 1994–2014), or by repositioning landscapes as art "under the sky," where the sky is the gateway to space, highlighting a relationship between nature and the cosmos (Tsong-Zung and Shiming 2016: 135–40).

The collection suggests that we are experiencing an existential crisis brought on by various factors in the advancement of technology, not least the possibility of becoming an interplanetary species, but also the rate of advancement of AI and other technologies that force us to confront the limits and possibilities of human and post-human bodies and minds. With the inclusion of works that are retrospectively repositioned in the context of human space exploration, including many works that primarily deal with other themes, the theme of interplanetariness is diluted; overshadowed by the broader theme of significant human-made threats to our species, including the threat of nuclear war and the climate crisis. It is sometimes difficult to separate these existential and physical crises from one another, and the overwhelming impression made by the exhibition is that interplanetary exploration may be driven primarily, and increasingly, by fear of the fate of the Earth. Indeed, Bertok's statement explicitly refers to a need to look beyond the vulnerable, "closed global system" of Earth at a time of climate crisis (Groys 2016: 13). Indeed, the climate crisis is a frequently cited motivation for space exploration, but the conflation of this and other crises, and the addressing of spaceflight only indirectly through the lens of human crises, limits the potential of this exhibition to explore the possibility of direct engagement with the environment of space.

Groys' exhibition helps audiences and artists to recognize unspoken connections between existing art practice and future spaceflight, highlighting the significance of space exploration for the human species, including human culture and identity. As such, it is an invitation for future artists to more explicitly and directly engage with these themes. By omission, however, it reveals the problems with producing and exhibiting art that more directly engages with the location of space and the conditions of spaceflight. Ultimately, the exhibition showcases art in the context of increasing interest in space exploration, not space art. This is understandable at a time when the art world is only beginning to identify the ways in which artists may engage more directly with spaceflight. The diversity of the work collected in Groys' exhibition, and the necessary omission of work that can only exist in space, evidences the challenges faced in representing this field in

Earth-bound exhibitions, and the difficulties encountered when attempting to define the emerging field of art practice that this book collectively categorizes as space art.

Toward a new definition of space art

There is, to date, no agreed terminology to describe the work that is covered by this book, nor a definition that fully aligns with the scope of the following chapters. This lack reflects the evolving nature of the field, as the arts seek to find their place in the emerging commercial space age. New fields of art practice associated with human spaceflight are identified variously as "space art" (Malina 1991; Raitt 2007), "astronautic art" (Woods 2022), "postgravity art" (Živadinov 2010) and "art satellites" (Bureaud 2021: 79). Authors offer overlapping definitions of these terms, with some agreement over the nature of the relationships between art practice and space, as well as some disagreement about the practices that are included in the field. Some reflect the state of practice at the time of their publication, in this quickly evolving field, and require updating to account for more recent developments and shifting attitude toward spaceflight. It is necessary to clearly define the field of practice that is explored in this book, not least to establish the author's remit, but also to offer practitioners the tools with which to define their own work and ambitions. A clear definition of the field, and the practices that are included and excluded from this definition, further helps to reveal where there are gaps in existing practice, and scope for artists to break new boundaries and establish new paradigms.

The term "space art" is appropriate to collectively describe a wide variety of practices that relate, in various ways, to space exploration. While the field remains small, an umbrella term may be adequate as a means of allowing artists to make connections and identify common concerns. However, as the spaceflight industry expands to offer a greater variety of opportunities for engagement, it is necessary to make more granular differentiation between different forms of engagement with space. Commentators employing the term "space art" have begun to find it necessary to list the different types of work that are encompassed by this term (Raitt 2007; Malina 1991; Woods 2019). These lists typically include a wide variety of practices that range from art located in space to art that is informed or inspired by spaceflight and the cosmos.

This lack of agreed terminology is largely due to the scarcity of practice that might constitute a field, as well as significant shifts in the goals and economies of the spaceflight industry. There have been limited opportunities for creative practice to take place in space, partly due to resistance from the spaceflight industry and misconceptions about the role that artists might place in space exploration, as will be explored further in the next chapter.

Also, a lack of recognition of this emerging field may in part be due to the difficulties exhibiting space art to audiences. A significant constraint on the aforementioned *Beyond the Globe* exhibition was, of course, that it was located on Earth, and could therefore only represent ambitions toward art that are directly sited in space. However, we are at a pivotal moment in the history of spaceflight, when artists are beginning to seek, and be given, opportunities for direct engagement with the environment of space. There now exists a sufficient range of existing practice that it is possible to begin to identify an emerging field of space-arts practice, and to distinguish this practice from Earth-based astroart.

At the forefront of this discussion is the question of location. Those concerned with direct engagement with space make explicit distinctions between "art sent into space" (Raitt 2007: 49) or "art-in-space" (Woods 2021b) and Earth-based art that engages with space indirectly, from Earth's surface. The siting of art has been the subject of significant debate throughout the twentieth century, as critical reflections on the ways in which the "white cube" gallery setting informs the readings of art led to, among other developments, the liberation of art from the constraints of the gallery setting. The potential siting or production of works in space is far more significant than a further step away from the white cube gallery setting and discourses surrounding the siting of art in which "context becomes content" (O'Doherty 1986: 15). Art-on-Earth and art-in-space, regardless of any themes that may unite them, are vastly different propositions. As this book will show, this is not merely a philosophical distinction, but rather, one that fundamentally impacts approaches to practice. Nonetheless it is not, at present, a clear-cut distinction. We are currently in a period in the development of the spaceflight industry when space is only accessible to a privileged few, and the majority of those who seek to experience space do so vicariously, through images and other documentation. For this reason, audiences may not witness spaceworks directly in the environment of space. Indeed, at least for the next few decades, even the artists themselves will not view their works of space art in the environment in which they are deployed. At this time in the evolution of space art, therefore, it is important to consider the location of the audience as well as the location of the work: one way to distinguish different categories of space art is to consider a work's location relative to the audience, or the location from which it is intended to be viewed. This is the approach taken by Malina (1991: 147) who has proposed categories that take account of the relative locations of work and audience: works located on Earth that are intended to be viewed from space; works located in space that are intended to be viewed from Earth; and works located in space that are intended to be viewed in space.

Malina (1991: 147) proposes two further categories that do not explicitly acknowledge the location of the viewer or the work: "art which makes use of new techniques, materials or sensory experiences generated as by-products of space exploration," and "art which expresses the new psychological experiences or new philosophical conceptions developed through space exploration." Although presented as mutually exclusive, it is feasible for there to be some overlap in these categories. Work that falls into these categories may be generated, located, and viewed on Earth, but this is not necessarily the case. It is possible, even likely, for work located in space to make use of new techniques or sensory experiences. As this book will later explore, weightlessness may be one such experience, generating post-gravity approaches to art on Earth and in space.

While this book is dedicated to the exploration of art that is situated in space, in order to test the limits of Malina's categorization it is helpful to identify some examples of Earth-based art that can be viewed from space. Spaceflight creates opportunities to view the Earth's surface and its artificial structures from great distances and new angles. This book will later explore the ways in which aerial photography has altered how we understand the landscapes that we inhabit (see Chapter 4). Birds-eye views and aerial photography that capture the landscape at orientations and scales that are not encountered in our everyday experiences on the ground, and ground-level and aerial views can offer two profoundly different experiences of large-scale land art.[6] Michael Heizer's *Double Negative* (1969–70), for example, is a 457-meter-long configuration of two straight trenches, 15.2 meters deep and 9.1 meters wide, sliced into the sandstone plateau of Mormon Mesa in the Nevada desert. This earthwork, remarks Todor Todorov (2014: 70) is best viewed in a satellite photograph, as it is only from a birds-eye view and distance that the scale of the work can be appreciated, and the "contrast between human intervention and natural formations" is apparent.

It is important to make the distinction between large artificial structures that can be viewed from space and those where viewing from space was the artist's intention, as per Malina's categorization, "works located on Earth that are intended to be viewed from space." There are significant examples of historical land art, such as Celtic chalk figures including the Cerne Abbas Giant, or other geoglyphs including the Nazca lines, that are of a scale and position that invite aerial viewing, but which were produced before this was a possibility. Human flight, and later spaceflight, has granted access to aerial views that were not previously accessible. As satellite technology has advanced and increased in resolution, it has become possible to capture images of increasingly smaller objects, meaning that we are entering an age in which a significant proportion of public artworks can be captured in

[6]Even where mediated experiences of land art have been opposed by artists, aerial documentation can become an integral part of the stewardship of the photographed works, documenting their inevitable erosion over time. Since 2012, aerial photographers have documented *Robert Smithson's Spiral Jetty* (1970) once or twice annually, recording the changing appearance of the work through the seasons and changing water levels. These aerial activities contribute to promotion and preservation of the work, and document work as an evolving, not static piece.

satellite photographs. Any work that can be viewed from space must be differentiated from works such as Pierre Comte's Pacific Triangle (1985), "inscribed within the coral atolls … of the Pacific Ocean" (Comte 1987: 18), or Eduardo Kac's *Lagoogleglyphs* (2009–present, see Case Study 6), where the artist has consciously set out to create a work with the intention that it be visible from space. This book will use the term *terrestrial space art* to describe works such as those by Comte and Kac that are located on Earth but intended to be viewed from space.

Other authors and artists have expanded Malina's categories of space art, including Arthur Woods, whose *Cosmic Dancer* is explored in Chapter 4. Woods (2019) offers a definition of "astronautical arts" with more granular distinctions than in the categories that Malina proposed three decades earlier. Woods' list of categories includes acknowledgment of the locations where the art has been created, acknowledging that commercial spaceflight now offers opportunities for the production of work in space. Woods adds to Malina's list: "art that is made in space," a category that prompts consideration for the ways in which the conditions of space impact the methods and materials of production (see Chapter 6). This Woods differentiates from "performance art, either in space or in a simulated micro-gravity environment," acknowledging that production might occur in real-time, as in performance art, and, further, that the simulated micro-gravity environment generates similar conditions for such performances as the environment of space. These categories are helpful in describing how painting becomes performance in Frank Pietronigro's *Drift Painting* (1998), performed onboard a NASA-facilitated reduced gravity flight (see Chapter 2). Video recordings of *Drift Painting* show the artist enclosed in a cube that frames the performance. He expels paint into the enclosed space as he tumbles weightlessly. The expelled paint becomes sculptural, adopting worm-like forms, and drifts through the air before falling to the floor as normogravity conditions are restored. Pietronigro (2004) argues that this expanded form of painting, that is only possible in microgravity, "eliminates the art object" so that the "commodity [is] replaced by ritual action."

Another helpful distinction proposed by Woods (2022) is that between works for "outer space" and works "designed for space habitats." The spaceflight community employs the terms "extravehicular" and "intravehicular" to describe these different environments. It is helpful to make the distinction between the environment of open space (the extravehicular environment) and the environment onboard a spacecraft (the intravehicular environment) for two main reasons: first, these environments pose very different physical challenges, with the extravehicular environment exposing artefacts to extreme temperatures and the vacuum of space, and the intravehicular environment having a controlled atmosphere that is, with the significant exception of gravitation, being comparable to that onboard an aircraft; second, these environments invite very different kinds of viewing.

The extravehicular environment is a space of vast distances, where large-scale spacework will likely be viewed by audiences who are located at great distances from the work itself. Annick Bureaud (2021: 79–81), describes large-scale works sited in orbit in the extravehicular environment as "art satellites," listing as examples Trevor Paglen's *Orbital Reflector* and Pierre Comte's *ARSAT.*[7] Such works contribute to the vast space-scape that is the night sky. "To create an artistic satellite is," writes Bureaud (2021: 80), "symbolically, to create something on the scale of the cosmos by placing a new celestial body within it."

By contrast, the intravehicular environment is an artificial environment—a human environment—in which humans will be in close quarters with any artwork that occupies that same space. There is potential, here, for a sense that the work and its audience are co-habiting in this enclosed, life-sustaining space. While Malina's exploration, and many of the proposed spaceworks proposed at around the same time, begin with an assumption that space art is necessarily large in scale, Woods recognizes that the interiors of spacecraft or space stations are more confined spaces, comparable in size to the architectural spaces that we occupy on Earth's surface. Wood's own *Cosmic Dancer* is intended for such a space, measuring only 35 × 35 × 40 cm in the form that was deployed on his behalf by cosmonauts onboard the Mir space station. Works intended for space habitats would be located and viewed from space, potentially at close quarters, with viewers inhabiting the same interior space as the work. Onboard space habitats, microgravity is the main condition that sets the environment apart from spaces on Earth's surface and, as Chapter 3 will show, requires a post-gravity approach that, as Maja Murnik (2016: 68) observes, "derives from entirely different foundations" to Earth-bound art practice.

Those foundations, as described by artists Dragan Živadinov, Dunja Zupančič and Miha Turši, and later philosopher Monica Puncer (2018), are "postgravity." Puncer (2018: 475) is critical of Malina's typology, intimating that in the years since Malina's definitions of space art were written, artists have developed more experiential appreciation for the environment of space, in part due to the activities that have taken place onboard zero-gravity flights. White's (2021: 28–9) exploration of the Overview Effect includes the observation that the experience of weightlessness deepens the space traveler's understanding of Earth gravity (1G), changing the human bodily experience so significantly that astronauts must relearn how their body

[7]Bureaud stresses the importance of making the distinction between "satellite art" and, borrowing a term from Comte (1987: 17), "art satellites." While Comte had previously employed the term "satellite art" in descriptions of his own work, Bureaud identifies a need to more explicitly highlight the potential for art to be launched into orbit, thus becoming an artificial satellite. Bureaud defines the former as "the use of existing satellites or their data for and within artworks" and the latter as "artwork visible from Earth of the creation of an artificial star."

operates (see Chapter 5). This postgravity experience is described wherever weightlessness is experienced, onboard reduced gravity aircraft as well as in space. The term, "postgravity art," like Wood's typology, represents the view that the distinction between Earth and space is less helpful than the distinction between normogravity and microgravity environments, since it is the gravitational environment that contributes most substantially to the difference between these two environments, at least to the extent that it is the distinction that has the potential to concretely affect approaches to art practice.

Postgravity art cannot be neatly contained within any of the subcategories of space art or astronautical arts proposed by Woods or Malina, largely because it has a conceptual, rather than concrete, definition. The term itself "indicates the overcoming of gravity," not just physically but intellectually (Puncer 2018: 480). Postgravity art may be located within the weightless environments of space or of zero-G flights, but may also encompass terrestrial works produced in response to an appreciation for the effects of weightlessness. Postgravity artists have either experienced weightlessness themselves, or observed its effects, and as a result have come to a profound understanding of the extent to which their previous practice has been constrained by gravity, or has taken for granted gravity's effects. Choreographer Kitsou Dubois (2005: 46) describes re-experiencing the human body and its movement potential in a weightless environment. During performances onboard CNES reduced gravity flights between 1993 and 2000, Dubois felt her body "merging with the empty space around." As a result of her experiences of weightlessness, Dubois came to understand human movement a "fostered lack of balance," and this in turn informed her approaches to teaching and choreographic human movement on Earth (Dubois 1994: 61; 63).

As this book will show, it is vital that art that directly engages with the environment of space is explored separately from art that depicts or otherwise involves data from space, as it is only by making this distinction that artists be given the space[8] to discover the extent to which this new environment can radically transform their practice. As progress is made toward the democratization of spaceflight, and space culturalization, it is important that artists are not left behind, dreaming of spaceflight while others experience it first-hand. Humans cannot be said to have colonized and culturalized space unless all of human culture is granted opportunities for involvement, and as the environment of space requires such radically different approaches to practice, vicarious involvement is not sufficient. Existing definitions, including those by Woods and Malina, are too accommodating in their willingness to list work informed by space alongside work that directly engages with the environment of space. Bureaud helpfully

[8]Pun intended.

stresses the importance of distinguishing art in space from art inspired by space, but her scope extends only to art satellites when other work might fall outside of that category. I propose a new definition of space art that extends only to art that directly engages with the conditions that are experienced in spaceflight, therefore distinct from astroart. It is art that can be described using this new definition that will be the focus of this book.

I propose that the term *spaceworks* may be used to describe artefacts that are located in and engage directly with the environment of space, including "art satellites" as well as artefacts located within orbital space stations, including art that is produced or deployed in these locations. The term stresses the significance of the location of the work, in space, encompassing some of the work that already falls within definitions of "space art" and "astronomic art," while excluding those practices that engage only indirectly with space, such as astroart and other solely Earth-based art practices that are informed by spaceflight and astronomy. Spaceworks are shaped by the same conditions as "postgravity art," in that the orbital environment is a microgravity environment, and those gravitational conditions have, as this book will show, significant influence on the characteristics of work in space, and the ways in which it is viewed. Figure 1.3 illustrates the main sites that will be the subject of discussion of this book and how they can be distinguished, through location, from terrestrial space art.

An apparent omission from many discussions of space art, and from Figure 1.3, is art that is sited on the surface of other planets. Paul Van Hoeydonck's *Fallen Astronaut* (1971) provides evidence of the potential for other worlds to provide new homes for artworks, including those that have been produced with those environments in mind, or as new locations for existing works that may be transported to other planets along with future colonizers and their descendants. While some such works are referenced in this book as they contribute to the larger context of art's engagement with space, this book consciously excludes such works from its definition of space art. The field of space architecture has established that the design of structures intended for space (such as space stations), and for the surfaces of other planets (as on the lunar or Martian surface), must be treated as two distinct fields of architectural practice. This distinction is made with good reason: the substantial differences in the environments at these locations. Conceptually, too, space is different from other planets. Outer space is a liminal space, where travelers are "detached from everyday life, removed from time, place and social structure" (Ormrod and Dickens 2019: 234). Space is between worlds, and even those that inhabit it do so temporarily. It is necessary for creative practitioners to make the distinction between space and other planets not only so that it is possible to more narrowly define the conditions that shape space art, but also to futureproof the term as humankind becomes more familiar with interplanetary habitation. Other planets are more comparable to Earth than to open space, and once humankind has colonized these planets they will be perceived less as celestial objects, and more as other Earths.

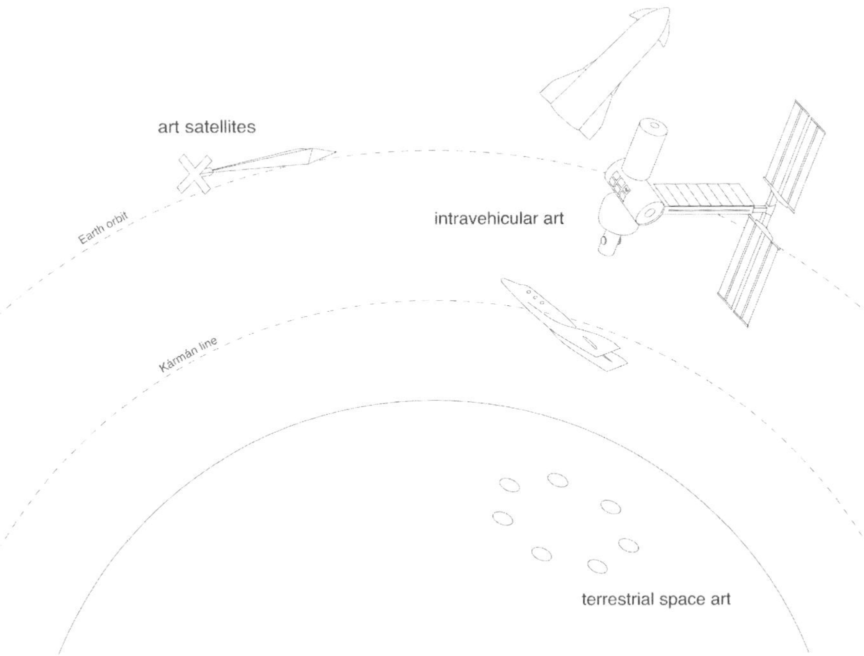

FIGURE 1.3 Diagram indicating the relative locations of art satellites, intravehicular art, and terrestrial space art. Terrestrial space art is located on the Earth's surface, visible from space (represented here by Pierre Comte's *Project Horus*); art satellites are spaceworks deployed in orbit in open space (represented here by Trevor Paglen's *Orbital Reflector*); intravehicular spaceworks are located onboard a vessel, which may move to different locations toward or beyond orbit. The vessels represented here are a sub-orbital spaceplane, orbital space station, and lunar rocket (Arthur Woods' *Cosmic Dancer* is represented here, framed by the window of the space station). Earth orbit is simplified in this diagram. Orbital trajectories are often elliptical and may be at different altitudes. The ISS is in low Earth orbit (LEO); most communication satellites are in geostationary orbit (GEO).

Case Study 1
Trevor Paglen, *Orbital Reflector*, 2018

Orbital Reflector is an artificial satellite, launched into orbit onboard a SpaceX rocket in 2018 as an "artistic gesture [that] could help to change the way we see our place in the world." Aiming to establish a place for artists in the occupation of near-Earth space, Paglen intended for his work to be the first non-military, non-scientific, non-commercial satellite into low-Earth orbit. The sculpture is a sword-like reflective polyethylene balloon, housed in a small satellite that was set to release the balloon in open space, allowing it to unfurl and inflate and become visible as a slow-moving star in the sky (Figure 1.4). The intended destiny of *Orbital Reflector* was for the work to orbit the Earth for a period of a few months, during which it would be visible from Earth, before burning up in the atmosphere. As a result of the 2019 government shutdown, which coincided with the planned deployment of Paglen's sculpture, engineers were not able to deploy the balloon as planned. The box containing the balloon continued to orbit the Earth, and the sculpture's destiny was unfulfilled.

Paglen's engagements with space arose from his experiences of observing and photographing the night sky. Paglen was interested in the apparently

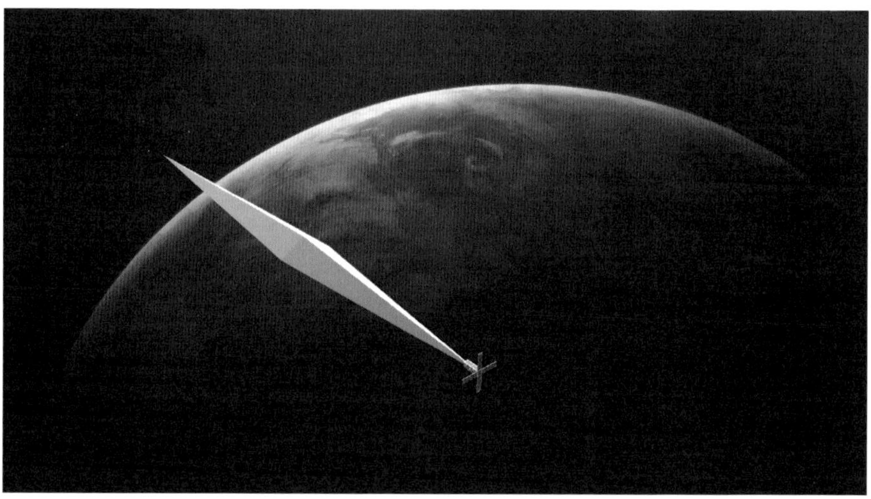

FIGURE 1.4 Trevor Paglen, Digital Rendering of *Orbital Reflector*, 2018. Once deployed, *Orbital Reflector* would have unfurled into a 100-foot-long diamond-shaped balloon coated with titanium oxide, visible to astronomers on Earth. The work was launched into space but never deployed, as a result of government shutdowns that affected federal communications and made it impossible to track the satellite. © Trevor Paglen. Courtesy of the Artist and Nevada Museum of Art.

empty space between visible celestial objects, where there are unseen spacecraft and military satellites conducting surveillance operations. This covert infrastructure appeared to be overtaking space that might otherwise have been used for objects that could more positively benefit humankind. These ideas yielded two spaceworks: *Orbital Reflector*, and *The Last Pictures*, which was launched onboard a communications satellite in 2012. The two works are very different in character and intention. Importantly for Paglen, both are "nonfunctional," meaning that they have no commercial, military or scientific purpose, existing solely to provoke thought. He hopes that his work will ask audiences to question what constitutes legitimate use of space.

Paglen is unsentimental about human uses of space. Although he engages in amateur astronomy and is professionally concerned with artificial use of space, he doesn't share the concerns of others who rally against nonfunctional satellites on the grounds that they will contribute to space debris. While he appreciates that such a characterization of artificial satellites serves as a useful allegory for human destruction of our own habitat, he suggests that it is the type, not the quantity, of artificial orbital objects that needs to be addressed. The "uselessness" of art satellites is meaningful in this context, provoking questions that would not be asked about "functional" satellites, or the motivations of the institutions that have launched them into space.

The development of *Orbital Reflector* was a lengthy process that could not progress without a large team of contributors and collaborators. Paglen describes building a community of people inside and outside of the spaceflight industry who were able to contribute the "advice and legitimacy" that was necessary before he could apply for funding to realize his vision for an art satellite. An important step on this journey was the production of images and prototypes that would make the project more concrete in the eyes of potential funders. A residency at Artpace in San Antonio gave Paglen access to an aircraft hanger, in which he was able to produce a full-sized prototype sculpture. This prototype, a mirrored sphere, was vastly different from the objects the Paglen eventually launched into space, but succeeded as a proof-of-concept and invited support from the Nevada Museum of Art, an institution with a history of commissioning and preserving land art, and that identified *Orbital Reflector* as an Earthwork.

The suggestion that *Orbital Reflector* could be considered an Earthwork, "made perfect sense," says Paglen. "An Earthwork is something that anybody can see ... and is very much part of the environment in which [they] see it." Earthworks draw audiences to a location for the purpose of viewing the work, and, once at that location, they prompt the audience to look around and appreciate the surrounding environment. Similarly, *Orbital Reflector* draws attention to space, being only one among many celestial objects visible in the night sky, and prompting audiences to ponder the many other objects, including other artificial satellites that are not visible to the naked eye but have contributed to an artificial cloak of objects that orbits the Earth. Like other Earthworks, Paglen's sculpture "has a very different relationship to an audience, to the planet, and with time" than works housed in a museum or gallery.

The work's relationship to time is of particular significance to Paglen:

> In space you can get lost in time in a way that it's not possible to get lost in time on Earth. In space there's no erosion, there's no geomorphic deposits, there's no volcanic activity. There's not much that effects the integrity of a structure that's in space. That cannot be said for anything that's on Earth. When you build machines and put them into space you're building monuments that can potentially be around for billions or even trillions of years ... You're working on cosmic timescales [and so] you can inhabit time in a different way

Without erosion, artificial satellites will outlive humankind's habitation of Earth, becoming a monument to civilization.

Paglen's other work to be launched into space, *The Last Pictures* (Figure 1.5), directly addresses humankind's relationship with time and the potential for artificial artefacts to outlive our species. The small crystalline wafer has etched on its surface one hundred images that represent human civilization, and hence the histories and contexts that led to the development of the work. Unlike *Orbital Reflector*, *The Last Pictures* is intended to be lost in space, "specifically it was meant to be lost in time. In space you can get lost in time in a way that it's not possible to get lost in time on Earth." The work is too small to be visible from Earth, and indeed, Paglen does not anticipate that it will ever be found or viewed by anyone in its orbital location. He stresses, however, that it would be unethical to assume that "people won't exist in the future" to rediscover the work. He aims to be "generous to the possibility" that it could one day be encountered by future spacefarers.

Orbital Reflector was lost in space unintentionally, but Paglen does not view this as a failure. He feels that there is "a kind of poetic justice" in the work's fate; a "perfect metaphor" for the way that the government had been stalled in its tracks, and in turn, how government has co-opted space for military gain at the expense of activities that would be more beneficial to humankind. The project had always been intended to provoke questions about "who gets to exercise power over planetary commons," and the failure of the government to maintain the conditions that would enable the deployment of Paglen's sculpture only serves to reinforce concerns about the balance of power on and beyond our planet. Paglen reflects on "early imaginations of space as a radically public space" and the ways in which young idealists enter the spaceflight industry, inspired by the words of Carl Sagan, only to find that the reality of the spaceflight industry is more committed to building weapons. Paglen describes how, during the

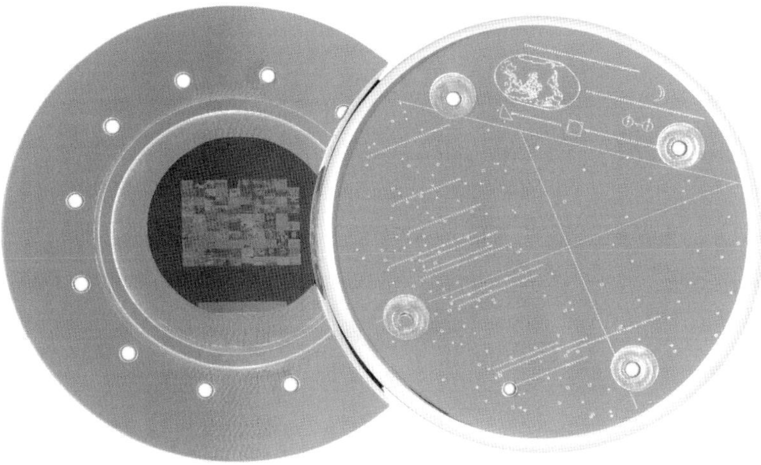

FIGURE 1.5 Trevor Paglen, *The Last Pictures Artifact*, etched gold-plated disk with etched silicon insert, 4 7/8" diam. × 3/8" thick, 2013. This disk is micro-etched with one hundred images that represent human life on Earth, including diagrams, illustrations, and photographs of landscapes shaped by human intervention, human gestures, and artificial artefacts. The work was affixed to a communications satellite, launched into geosynchronous orbit, and then later intentionally lost in space. © Trevor Paglen. Courtesy of the Artist, Altman Siegel, San Francisco and Pace Gallery.

project's development, he found support from spaceflight industry insiders who found hope in his "art-for-art's-sake" proposal because it represented to them the ethical, social, and poetic ideals that the industry has lost.

In the few years since the launch of *Orbital Reflector*, there has been significant change in the spaceflight industry. At the time of writing, there is a "streamlined workflow" for space payloads that did not exist when Paglen planned the deployment of his art satellite. While this made the realization of the project complex, it also meant that Paglen's ambitions were not constrained by the payload parameters and expectations that have shaped the work of the other artists interviewed in this book. One significant challenge for Paglen was the financial cost of launching objects into space. This included the cost of launch as well as other peripheral costs like insurance, which were affected by insurers' uncertainty about how to value a nonfunctional payload.

Trevor Paglen was interviewed on February 28, 2023.

Case Study 2
Yasmine Meroz and Liat Segal, *Impossible Object*, 2022

At inception, Meroz and Segal's ambition for *Impossible Object* was to create work that can only exist in space; "to do something that cannot happen here [on Earth]." The work features interconnected brass rods and tubes, through which water flows. The water pooling in spheres on its surface forms an essential part of the sculpture's overall shape. Unlike on Earth, where liquid pools to adopt the shape of a vessel, in microgravity it adopts shapes independently of other objects, becoming a sculptural form in its own right (see Chapter 6). As a result of this behavior, Meroz and Segal's sculpture can only exist in this form in the weightless environment. In microgravity, the work becomes "a dynamic three-dimensional liquid composition," in which the pooling water appears to have as much substance and integrity as the metal frame to which it adheres (see Figure 1.6). The work achieved its potential in the microgravity environment of the ISS in April 2022, where it was deployed by Israeli astronaut, Eytan Stibbe. Video recordings of the work in space show water pooling in undulating spheres that migrate along the brass rods, occasionally releasing small droplets of water into the surrounding environment.

Segal and Meroz' initial starting point was to identify that "physics is the medium" for this work. Having previously worked at the intersection of art and technology, Segal is wary of the risks of engaging with new technologies and conditions. She has found that, oftentimes, "when you use technology in art, it becomes a gimmick." This is because, she suggests, the work often becomes "more about the technology than it is about the art or the story." Her solution has been to treat technology as her medium, not her subject. The risk of gimmickry could also endanger art which attempts to engage with space. Meroz and Segal therefore identified at the outset that in order to engage with the environment of space, it should be their medium, not their subject.

The mode of deployment added further unearthly dimensions to the work. Sending the work into space meant relying on Eytan Stibbe to construct it on behalf of the artists. This necessitated rehearsals on Earth, during which they supported the astronaut during his training to build the sculpture and connect it to a water pouch. The process of construction, as rehearsed on Earth, differed significantly from what was possible in space. Video recordings of the work's deployment on the ISS document how difficult the challenge of construction was in the weightless environment. These difficulties arose not only as a result of the objects threatening to float away from the astronaut, but also the impairment of fine motor skills that astronauts experience before their bodies become accustomed to unfamiliar

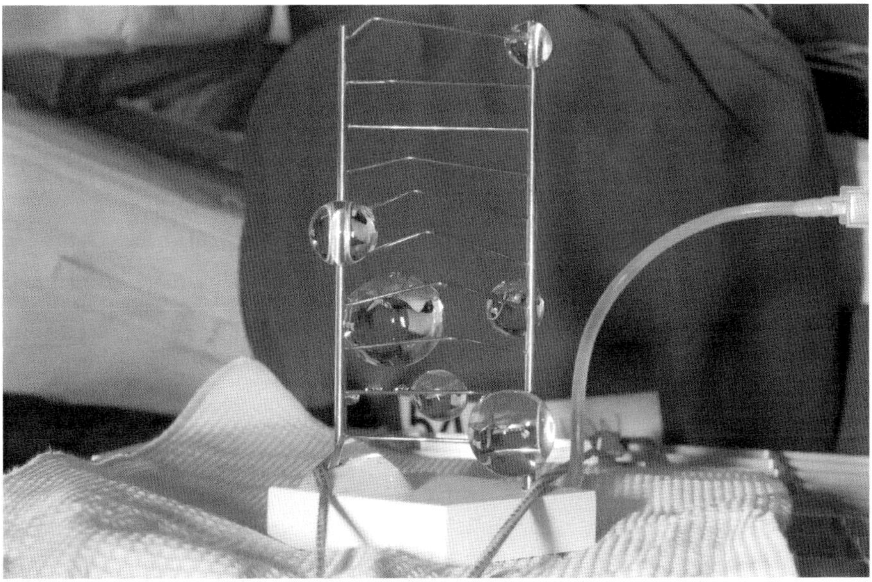

FIGURE 1.6 Yasmine Meroz and Liat Segal, *Impossible Object*, brass, aluminum, silicone, paper, water, 10 × 10 × 22 cm, 2022. Onboard the International Space Station in April 2022, water clings to the brass frame, forming large spheres. The points of adhesion, where the water connects to the brass rods, appear delicate, as if they could be easily disrupted by human interaction or other forces. As a result, the work gives an impression of fragility and impermanence. Videos and images onboard the ISS were taken by Eytan Stibbe, as part of Rakia Art Mission (Ramon Foundation). © Yasmine Meroz and Liat Segal

gravity conditions (Holden et al. 2022). The astronaut was also responsible for video-recording the process of construction and the completed sculpture, capturing astronauts in "playful" interactions with the work. Segal describes the pleasure and importance of seeing, via video recording, the ways that the astronauts engaged with the work in-situ. For Meroz, this experience was deeply emotional. Parallels between the playful experiments that the artists conducted on Earth and the ways that the astronauts responded to the work in space reinforced a connection between herself and the astronauts.

For the artists, it is impossible to disentangle the sculpture from the surrounding events and experiences. This, Meroz suggests, extends the ways in which the work is different from anything that could be created on Earth. Since the sculpture only exists in space, and is therefore experienced by most audiences via a video, the work exists beyond merely the physical sculpture in its setting. Meroz refers to the work as a "story" that has no definitive beginning or end, existing, both physically and temporally, beyond the location and moment of the sculpture on the ISS. The story includes the performance of unboxing and construction by the astronaut, and his act of documenting that process. Segal suggests that "even our work in the studio is part of the story," and, in particular, their efforts to engage with "the unknown." This whole process, observes Meroz, is "fundamentally different" from the way that artwork is created and accessed on Earth.

Before deployment, the brass frame of *Impossible Object* was collapsed and stowed in a box, along with other components (see Figure 1.7). The artists describe this form of presentation as an important part of the work. The box, and the sculpture's clean, white pedestal, are minimalist in design, contrasting with the cluttered surfaces of the cabin of the ISS, and this contrast between the box and its environment helps to reinforce the distinction between the scientific and artistic activities that might take place in this space. Efforts to ensure that the work does not resemble scientific equipment emphasize the different ways that the artists expect their audiences to receive and interact with their work as a cultural, rather than scientific object.

Both Segal and Meroz have scientific backgrounds: Segal studied computer science and biology and was a researcher in Machine Learning before shifting her trajectory toward art; Meroz has a background in physics, and is a Principal Investigator at the Faculty of Life Sciences at Tel Aviv University, studying the physical concepts underlying computational and behavioral processes in plants. These scientific backgrounds, and the influence of their past scientific research, is very apparent in the way that they describe the development phase that led to the creation of *Impossible Object*. Knowledge of the way that plants interact with water has had

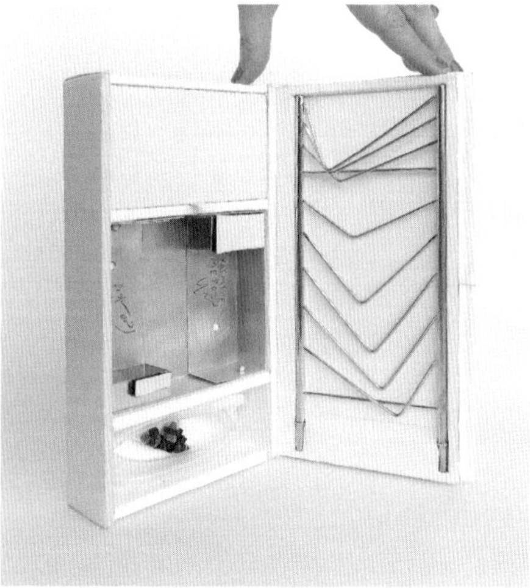

FIGURE 1.7 *Impossible Object*, boxed payload. The brass structure is collapsed for stowage. In addition to the objects contained within the box, the work requires a drinking pouch and water, already available onboard the ISS. © Yasmine Meroz and Liat Segal

significant impact on the direction of the project, and has, to some extent, substituted for direct experience with microgravity. Her knowledge of plant behaviors equips Meroz with an understanding of the forces that govern the behavior of water. Water, she notes, is shaped by several forces, including gravity, adhesion forces, and water tension. When water exists in "very, very small drops ... then gravity is negligible." Plants can "use adhesion forces and water tension" in order to overcome gravity, and are able to transport water upward, for example, from their roots to their leaves. This knowledge gives insights into the way that water can behave when gravity is not the dominant force that shapes its behavior. Similarities can be seen in the way that water is shaped into perfect spheres at a microscopic level on Earth, observable on plants in morning dew, and the larger spheres that the water adopts around *Impossible Object* onboard the ISS.

Despite their experience in science and technology, the artists were not fully able to predict the way that their work would behave in microgravity. While a lifetime of experience of water on Earth gives us "strong intuition about how water can behave," for artists "this intuition [about how materials will behave] is usually gained by practice." Without opportunities to experiment in microgravity, it was necessary to gain this knowledge through observation of experiments conducted by others. Segal describes watching many hours of videos filmed onboard the ISS, in particular those featuring astronaut Don Pettit's experiments with water. These videos, available publicly via NASA's online archive and YouTube,[9] demonstrate the potential for water to adopt shapes, or to envelop objects. Segal recalls videos in which water is shaped into lenses and sheets, taking forms that are akin to soap films and bubbles on Earth. The videos also reinforced the artists' understanding that the way water would interact with the material chosen for their sculpture would depend on whether that material is hydrophilic or hydrophobic. If a material is hydrophilic, "the water would cover it like an envelope." At the other end of the spectrum a hydrophobic material would disconnect from the water, causing the water to float away.

These new and existing understandings of the behavior of water on Earth and in space gave them a general sense of how their sculpture might behave, but they did not know definitively what might happen in space. There was a spectrum of possibilities, and they knew that where the water's behavior fell within that spectrum would depend on the materials used for the frame as well as on the actual activation of the sculpture by the astronauts. They conducted numerous experiments in their studio, in attempts to gauge the extent to which water might cling to their brass frame. While these experiments were helpful, there was always an "unknown." They describe

[9]The best of these videos are collected in the YouTube playlist, 'Don Pettit's experiments in space,' available at https://youtube.com/playlist?list=PL7E4B5CB811F0EFEC&feature=shared

the eventual deployment in space as a further step in these experiments, as, even at the time of the work's final realization, they were surprised by the shapes that were adopted by the water around the frame. Having predicted that the water would, to a greater or lesser degree, envelop the frame, the water's form was "more spheric than we had guessed [*sic*]" (see Figure 1.8).

Another unknown was the size of the connector through which the sculpture would interface with the water pouches that were already onboard the ISS. The artists describe the laborious process of attempting to uncover this information. Without definitive measurements, it was necessary for Meroz and Segal to adapt their design to accommodate a range of possible connector sizes, and to include within the payload a backup connector of a different size. This difficulty in accessing necessary information was

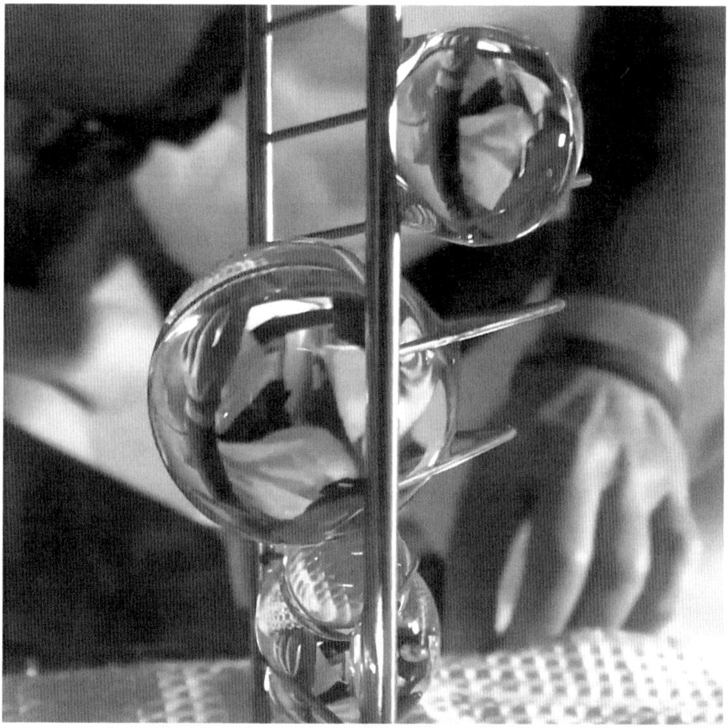

FIGURE 1.8 *Impossible Object*, close-up. Before deployment, Meroz and Segal were uncertain about the extent to which water would cling to their brass frame in microgravity, and conducted a number of experiments and observations on Earth with a view to being able to predict this behavior. They had anticipated that the water might envelop the structure, and were pleasantly surprised to observe the near-perfect sphere's that the water adopted when the work was deployed. Videos and images onboard the ISS were taken by Eytan Stibbe, as part of Rakia Art Mission (Ramon Foundation). © Yasmine Meroz and Liat Segal

not, they stress, the result of any reluctance by others to support their project, but, rather, the result of bureaucracy and the long and complex chain of individuals and organizations that lay between themselves and the information.

Segal and Meroz are hopeful that barriers like these do not deter other artists from seeking opportunities to produce work for space. "We would like to see more opportunities for art in space," says Segal, but she observes that there are numerous challenges for any artist who shares their ambition. "A very big barrier right now is money," says Meroz. The cost of sending payloads to space is prohibitive, meaning that any art initiatives are "very dependent on patrons." For Meroz and Segal, the opportunity to create *Impossible Object* arose when the Ramon Foundation approached Israeli researchers and artists to propose projects that could be deployed during the Rakia mission. Without support from third parties like these, artists would have no opportunities to work in this emerging field. "It's not just about providing funds," says Meroz, "it's also about opening channels." Calls for participation, initiatives that invite ideas beyond science and technology, together with a general willingness to consider the role of the arts in space are all necessary for artists like Segal and Meroz to produce further impossible works.

Yasmine Meroz and Liat Segal interview conducted February 9, 2023.

The work "Impossible Object" was supported by Mifal Hapais, Asylum Arts, the European Union's Horizon 2020 Research and Innovation Program (GrowBot), and The George S. Wise Faculty of Life Science, Tel Aviv University. Videos and images onboard the ISS were taken by Eytan Stibbe, as part of Rakia Art Mission (Ramon Foundation). Rakia Art Mission curator: Udi Edelman.

2

Science First, Art Later

Art in the space sector

Spaceflight is possible thanks to remarkable achievements in science and technology. Scientists and engineers have developed the technologies that have carried humans to space and allow us to survive in its inhospitable environment. It is perhaps an inevitable consequence of this history that "space missions assign a primary role to science and technology and allocate a peripheral place to more elusive and fluid concepts like culture and history" (Noble 2022: 71). Missions must prioritize safety and efficiency, and do so at the expense of "more human-centric considerations" (Sharma 2020). The expansion of spaceflight industries in recent decades, alongside increased outreach activity from governmental space organizations including publicly available recorded and live footage of life onboard the ISS, has led to more widespread familiarity with the environments encountered during spaceflight. Artists are increasingly empowered to imagine those environments and how they might engage with them. Space is no longer viewed as an environment exclusively accessible to trained astronauts or those with specialist aeronautical, technical, or scientific expertise. Speaking on the development of the proposed space station Orbital Reef, Brent Sherwood of Blue Origin observes that in a commercial world "advanced capabilities get normalized and become a part of everyday life" (Faust 2022). When technologies are commercialized, they become accessible to laypersons, requiring less specialist expertise from their users. In this current era of commercial spaceflight, passengers and others indirectly contributing to spaceflight will increasingly be able to operate without an advanced understanding of the science or technology that enables space exploration.

Artists attempting to engage with spaceflight have more opportunities than ever before, but still face barriers. While the spaceflight industry has begun to recognize the need for a relationship with the arts, there is still widespread lack of understanding about how and why artists could contribute value to their missions. NASA proudly asserts a longstanding relationship with the arts, but scratching the surface reveals that their space arts initiatives have invited participation from hobbyists rather than professional artists, and mostly as an outreach activity, excluding the possibility of direct involvement with the environment of space. Planning to deploy a small-scale sculpture in space in the early 1990s, sculptor Arthur Woods (2005) was informed by NASA that there was, at the time, no program for the launch of non-scientific payloads on the space shuttle. Woods had more luck with the Russian space agency Roscosmos, and his sculpture, *Cosmic Dancer*, was ultimately deployed onboard the Mir space station in 1993 (see Chapter 4). This success was not as a result of any Russian drive toward engagement with creative communities, but rather, a need to fund the Russian Space program in part through commercial investment. Much has changed since NASA rejected Woods' proposal, but governmental space organizations are still grappling with the question of how to benefit from a relationship with artists. Even where they have involved the arts there is a sense that this art is only of value if it promotes some scientific or public-relations function.

When the Beagle 2 rover landed on Mars in 2003, it carried with it a spot painting by British artist Damien Hirst, intended to serve as a calibration target for onboard cameras (Gérin 2003: 37). The work is a small aluminum sheet on which are painted concentric rings of dots, including white and black dots to aid onboard instruments to calibrate their white and black values, and a range of yellow and red hues representing iron at various grades of oxidation, to which Martian soil could be compared. Hirst added "non-functioning" green and blue dots to represent Earth (Weddigen 2018: 339). The work faced criticism for including "art [only as] a surplus gadget attached to a scientific machine" (Weddigen 2018: 340). The characterization of the piece as "art" fueled these criticisms, highlighting that the work's artistic function was secondary to its principal, scientific function. Commentary, including comments by Hirst himself, place function and aesthetics in opposition. Hirst (cited in Radford 1999) chose "function over aesthetics" not only suggesting that aesthetics must be compromised for functionality, but also reinforcing the perception that there is binary opposition between science and art, including the narrow perception of science as a field that is principally concerned with function and art as a field principally concerned with aesthetics. Commentary by Tim Radford (1999) in *The Guardian* recorded that "no money is changing hands; the painting is seen as of mutual benefit to science and art," with the implication that art, when produced for the sake of advancing science, should be a philanthropic activity when carried out in service to science.

When art is launched into space, the ways in which it is characterized by commentators reveals much about the complexities of the relationship between art and science in the spaceflight industry, and about how the arts may be (mis)understood by those working in spaceflight. JAXA, the Japanese Aerospace Exploration Agency, has supported arts experiments onboard the

ISS including Takuro Osaka's *Spiral Top* (2009, see Figures 2.1 and 2.2) and *Spiral Top II* (2017). Osaka refers to this work as "space art," and JAXA describes it as an "art project." Osaka (2004: 274) locates the work in the broader context of emotional and aesthetic responses to spaceflight. NASA's characterization of Osaka's work is vastly different. NASA's image library does include images of the work, but nowhere in the image descriptions do they refer to art or related terms, rather, they characterize it as a scientific experiment "to record the motion of a spinning top," indicating that the purpose of this work is to produce quantitative data. The lack of reference to art or related terms means that any stakeholder searching NASA's archives using these search terms would not find a record of this work, nor would they find records of this work by searching for the artist's name, which is also omitted from the library. These omissions are a failure to represent the arts and artists and to recognize their contributions to space exploration.

We can learn much about the barriers to artistic engagement with spaceflight, and the progress made toward overcoming them, through the

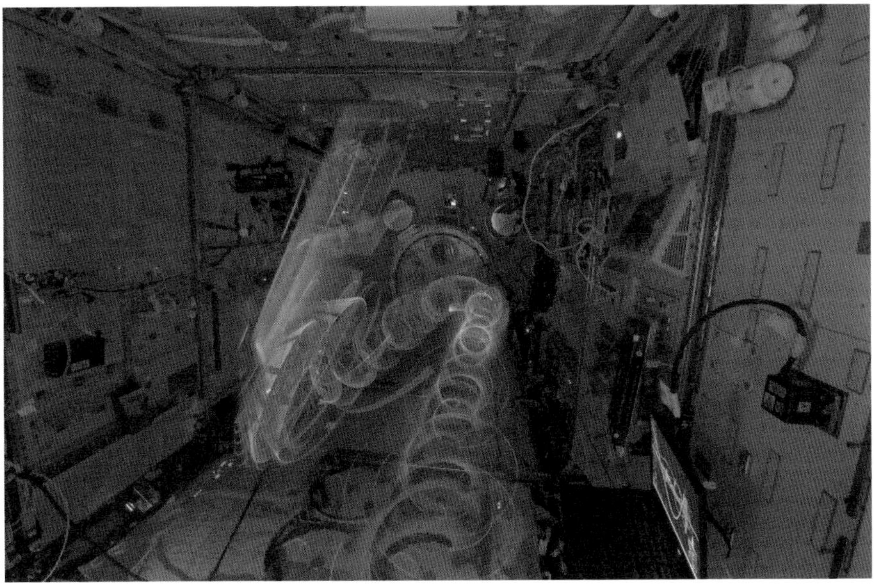

FIGURE 2.1 Takuro Osaka, *Spiral Top*, 2009, shown illuminated and spinning onboard the ISS in 2014. When describing an early prototype, *Comic Spiral Top*, Osaka (2004: 274) invites an emotional and aesthetic response to the work, suggesting that "while enjoying its beauty, people in outer space may think of the connection between the double helix structure of DNA and the helical movement of the universe." While Osaka and JAXA both refer to Osaka's work as art, NASA makes no reference to art or to the name of the artist in their archives, instead characterizing the work as an "experiment ... to record the motion of a spinning top onboard the International Space Station." Image © NASA

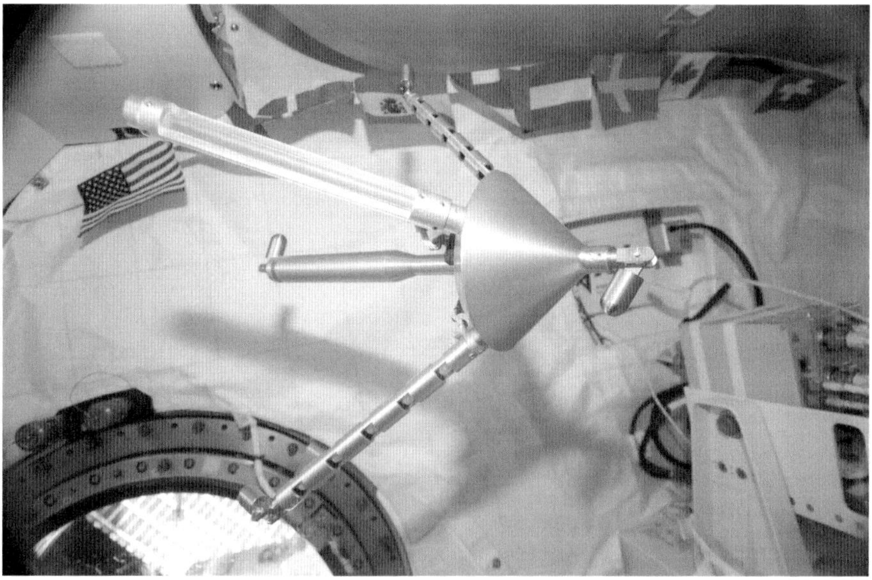

FIGURE 2.2 Takuro Osaka, *Spiral Top*, 2009, pictured in 2014. Here the *Spiral Top* device is shown dormant prior to demonstration. During demonstration, the work is illuminated by an array of colored LEDs on its limbs, and it is manually spun before being released into the cabin to continue on its spiral path. Weights are attached to its limbs in order to unbalance the spiral motion and create irregular light patterns. Image © NASA

space policies of spacefaring nations. Most of these policies are ambitious in scope, recognizing that space exploration will yield new insights that transform life on Earth and contribute significantly to national economies. All are united in recognizing space as a site for scientific research, and most recognize potential for collaboration across disciplines or sectors. Many reveal a significant gap between stated goals and any progress made toward realizing those goals in practice. Others neglect to acknowledge the arts altogether. The UK Space Agency (UKSA) provides one such example (gov.uk 2022). Outlining their vision for the UK spaceflight industry in their National Space Strategy, the UKSA states that it "will build one of the most innovative and attractive space economies in the world." In its stated goal to "lead pioneering scientific discovery and inspire the nation" it is implicit that inspiration can only come from scientific and not from creative engagement with space. They identify opportunities for the UK's science, business, financial, and legal sectors, but nowhere does the strategy mention the creative industries. They identify the need to "support research that ... allows us to live ... richer lives and answers the biggest questions about our place in the universe" without acknowledgment of the role that the arts could play in achieving that goal. The UK's failure to acknowledge opportunities for the inclusion of the arts is despite the ESA's inclusion of many UK-based arts organizations in consultation and arts initiatives. In the UK and elsewhere, initiatives in support of the arts have emerged tangentially to government organizations, but are rarely core to government-initiated missions. The most significant contributions to the future of arts' involvement with spaceflight have arisen in unlikely locations, where the spaceflight industry is still new and developing. The emergence of fledgling space organizations in developing nations therefore creates further opportunities for more widespread involvement from the arts, as well as opportunities to influence government space policies as they are being formulated.

NASA has explicitly supported the arts, and proudly cites a long history of initiatives that invite artists to find inspiration in their activities (Ulrich 2007). The examples identified in their celebration of the NASA Art Program are all representative, depicting NASA's real and imaginary achievements in oil and watercolor paintings of astronauts and spacecraft. They describe the role that these images play as a "record of this nation's program of space exploration," to "supplement" existing records and fill gaps identified in their archives, thereby positioning these images alongside the other visual and verbal records produced by astronauts and other actors in NASA operations. The Art Program has been scaled back in recent years but continues to exist alongside other initiatives that take a slightly different view of the value of art for NASA. Today, NASA primarily supports arts as an outreach activity. Many of these invite submissions from children or other non-professional artists, as a way of engaging these stakeholders indirectly in their activities. These initiatives include, for example, the Space Place Art Challenge, where "art" is equated with "drawing" and school children are invited to "draw a space-related situation each month." In many such initiatives, art is employed as a tool to make space more accessible to audiences who do not yet understand space science.

Art continues to be viewed as an outreach activity when it targets adult audiences. The Studio at NASA's Jet Propulsion Laboratory (JPL) is a team that includes artists and designers. Its recent output has included *Visions of the Future*, a series of posters that imagine distant planets as tourist destinations, and *Orbit*, a sound installation that interprets location data from passing satellites as sound. In this installation, the passage of satellites is interpreted as "naturalistic ... electronic [and] synthesized sounds," played on a spatial audio system so that the sound "moves along the satellite's trajectory." The soundscape is broadcast in a metal, spiral-shaped pavilion at The Huntington Botanical Gardens, where audiences are invited to experience the installation as an "educational experience," that enables the public to "understand that ... satellites move above them" (NASA 2016). While the Studio's works are varied, they all take space as a subject, as opposed to a site for art practice. The Studio's declared objectives—"helping scientists and engineers imagine the future, and giving people a sense of awe about the universe"—suggest that they imagine two different audiences for their projects: scientists and the general public, although their output to date appears to target mainly the latter.

NASA's Hubble Heritage Project "uses data to create striking images for the public." Those involved in the production of these images acknowledge a conscious disconnect between the scientific and artistic aims of the initiative. The project coordinator, Jayanne English, describes the goal of the project as being "to disseminate science-rich images that caught the public's eye because of their aesthetic—they did not need to be newsworthy but did need to represent the scientific results of our colleagues" (Mir 2017: 49). English describes delivering basic color and composition workshops to the Hubble Heritage Team, training them to apply color to images captured by the telescope to make them more visually appealing to a lay audience. She describes the result as "eye-candy" and acknowledges that such initiatives are "successful as public outreach" though "their intention is not to make art" (ibid., 50). Artist Aleksandra Mir (2017: 53–4) observes that this project is at odds with the purpose of art, because "the purpose of the Hubble images is to sell the science of the telescope, not to advance aesthetics or critical thought." She records that the Hubble team frequently discards images that are ambiguous or unclear, or otherwise considered to be imperfect, a revelation that will horrify many artists, who might find opportunities for creative or intellectual engagement with these images partly because of their ambiguity.

By focusing on the role of the arts as promotion and outreach, and by equating art with representative or decorative images, NASA denies the many other roles that the arts can play, and dictates the kind of art that must be produced in this context. David Raitt (2007: 49) outlines problems arising from the perception of space art as an outreach activity, intended primarily to inform the public and increase public awareness of the spaceflight industry and its discoveries. There is a belief that the purpose of this art is to "depict and visualize the cosmos" and as a result, "space art is rarely considered to be 'serious' art," but rather, is employed as illustration. This kind of astroart, that is "most often found in technical museums and planetariums," is what many in the spaceflight industry envision when they are asked to imagine how artists might engage with their sector. This view is pervasive in much of the world, particularly the USA, where NASA's arts and culture initiatives reinforce unhelpful preconceptions. One reason that these perceptions are so difficult to overcome is that many arts and culture initiatives are established by scientists and educators with an informal interest in art, as opposed to practicing contemporary artists.

Where NASA has real opportunities to invite participation from artists it falls short. NASA has a significant stake in the ISS, which, until the recent launch of China's Tiangong space station, was the only space habitat that could host long-term creative engagement with the environment of space. One of NASA's (2022: 22) five mission goals for the ISS is to "inspire humankind." Those working in the arts might identify this as a call to consider the creative and philosophical opportunities that spaceflight can provide, and yet, when elaborating on this aim, NASA refers only to STEM (science, technology, engineering, and math) and not to other sectors. Their stated ambition to "stimulate diverse groups ... interest in, and curiosity about, space" is located alongside an explicit assumption that this diversity will extend only to STEM students and a STEM workforce.

Europe has made more progress than the USA in recognizing opportunities for the arts in space. Early attempts by the ESA to engage with the arts were, wrote Annie Gérin in 2003 (p. 37) "somewhat gimmicky." Recognizing a need to address this, the ESA took action. In 2005, it commissioned London arts organization The Arts Catalyst (2005) to identify opportunities for "cultural utilization" of the ISS. The report recommended changes to policy in favor of permanent inclusion of cultural activities that would reflect the "cultural diversity of Europe and the rest of the world." The report stresses the need for inclusion of curators as well as artists, who are employed from outside the ESA, acknowledging that, at that time, there was not yet "appropriate cultural expertise within ESA" to manage a program of cultural activity. The Arts Catalyst was ideally placed to make these recommendations, having demonstrated the potential for engagement with weightlessness through its Microgravity Interdisciplinary Research program (2000–4). The program granted artists access to weightlessness onboard reduced-gravity flights in Russia and France, and yielded detailed philosophical observations and reflections that were very different in character from the insights published by professional astronauts.

In 2011, the ESA established a Topical Team Arts and Sciences (ETTAS), which included artists, curators, and philosophers as well as members from the ESA and other space exploration organizations, to explore the possibilities for cooperation with the arts. The group's initial findings include observations about the differences between art/sci and art/tech collaborations that already take place within the arts and spaceflight sectors. They observed that "research-oriented multidisciplinary" collaboration already widely takes place in the arts. Artists are, therefore, familiar with approaches to cross-disciplinary collaboration and practice that can be located within research contexts and aligns with the goals of research organizations, including scientific and technical research bodies. By contrast, scientists in the group had experience of commissioning, rather than collaborating with artists, in the context of a need to achieve outreach goals in communication, education, and marketing (Pell et al. 2014: 182). As a result of these different experiences, artists and scientists have varying expectations about how the arts and sciences may work together, with artists envisioning sharing common research goals with their collaborators, while scientists imagine employing artists primarily in the role of interpreters or communicators whose work is secondary to their own scientific research findings. These observations highlight a problem that underpins barriers to artists' involvement in spaceflight, but also demonstrate progress toward acknowledging these barriers and why they exist.

The ETTAS concluded that opportunities for the inclusion of artists could be implemented without significant alteration to existing infrastructures, in three ways: calls for participation, residencies, and fellowships (Pell et al. 2014: 183). It proposed that these efforts could yield artworks informed or inspired by ESA activities, "artworks that use infrastructure as agent," and "artworks that contribute to ESA projects" (ibid., 186). These categories could include work created and deployed on Earth as well as in space, and were informed by examples of work by artists working in a range of media and contexts including artists' engagements with weightlessness onboard reduced-gravity flights. In order to ensure the quality and appropriateness of these activities, the team concluded that it would be necessary to form a jury of curators, artists, and scientists for the purpose of reviewing artists' proposals (ibid., 185). The inclusion of both arts and space industry representatives on this jury would ensure the value of any outcomes to both the space exploration and arts sectors, and that the selection would reflect the diversity of arts practices.

The ESA now operates an artist-in-residence program that, vitally, extends to non-representational arts including sound art, interactive art, and choreography. Aoife van Linden Tol, the first resident, experienced opportunities to operate directly inside the ESA, witnessing discoveries as they took place. She took up residence in an office in the European Space Research and Technology Centre (ESTEC) facility in Holland, and embedded in that location she worked alongside the ESA, attended the same lectures and training as ESA employees, and used her new knowledge to inform the works that she produced during her residency. She describes inviting ESA scientists and engineers to collaborate with her during her creative process, and shared with them her methods of production, allowing them to shape her ideas at the point of conception (van Linden Tol 2017). The resulting work, *Star Storm*, is a series of controlled explosions that mimic the processes of the stars. In its current form, the ESA residency program supports collaborations between artists and scientists working in fields related to spaceflight. This program does, however, have limitations. It is Earth-based, supporting indirect, intellectual rather than actual engagement with space. Further, it contributes to a perpetuation of the view that science precedes art, describing the role of the artist in these art/science collaborations as to learn about science, while the benefit to the scientists is described as an opportunity to "experience ... working with an artist" (O'Flaherty and Isaak 2021: 6). In this way, science is privileged in these residencies, where collaborations are founded on the idea that artists have much to learn from the sciences, but not vice versa.

CNES, the French space agency, similarly hosts a range of initiatives including artists' residencies, but goes a step further than the ESA in enabling direct engagement with the environment of space. In 2000, CNES established *Observatoire de l'Espace*, to enable cultural engagements with space. These have included the *Résidence ZERO-G* program, in which artists are enabled to carry out activities onboard parabolic flights (see Chapter 3); in the future it will include opportunities for deployment of spaceworks in orbit. The *ArtSatel* program, launched in 2024, invites artists to propose works for deployment onboard a space capsule in orbit, with an anticipated deployment date in 2026 (CNES 2024a). The selected artist or collective will not be able to accompany their work into orbit, but rather will need to produce work that is deployed mechanically. The progam represents a significant step beyond the philosophical and hypothetical engagements with space that have been enabled by other initiatives, toward a future in which direct physical engagements with the environment of space are supported by government agencies.

JAXA (2016: 1), likewise, has dedicated resources to exploring the environment of space as a site for art practice. In 1996, the National Space Development Agency of Japan (NASDA) launched an initiative to explore cultural and social sciences approaches to space exploration, inviting research in fields including philosophy, religion, sociology, literature, and art. Having established collaborative research agreements with arts institutions, Tokyo University of Arts and Kyoto City University of Arts, in 1996 and 2000 respectively, they later established, in 2012, an art subcommittee. In contrast to the ESA's arts initiatives, largely contained within Earth's atmosphere, NASDA and JAXA have encouraged direct engagement with the environment of space. JAXA (2016: 3–4) asserts that the ISS "should be a place not only for pursuing scientific outcomes, but also for tangible and intangible contributions to all of humankind." This recognition of the potential for space, and more specifically the ISS, as a site for art practice established the conditions that would allow, for example, artist and academic Takuro Osaka to create several works for deployment onboard the ISS in 2008–14, some of which are explored in this book. As illustrated by Osaka's body of work, JAXA encourages art research that promotes a culture of communication between the arts and sciences, yielding outcomes that draw from, and benefit both the arts and science communities.

JAXA (2016) presents artistic experiments as a natural continuation of scientific experiments that have been carried out previously, which are typically intended as initial investigations into behaviors and properties that might have some as-yet-unknown future application (see Chapter 6). They are, however, keen to stress the differences between scientific and art experiments, observing that their sponsored "art experiments seemed to be elusive, nonsensical, and pleasantly optimistic compared to scientific experiments, which verify hypothesis by conducting space experiments." This difference, they found, empowered astronauts to operate in new ways, with freedoms that they had not previously experienced. In particular, JAXA reported that astronauts felt empowered to exercise creativity, injecting their own ideas when deploying the work and embracing opportunities for productive play.

The most ambitious and inclusive government-supported space arts initiative that has taken place in recent years is the Israel Space Agency's Rakia mission, supported in part by funding from Israel's national lottery. In 2022, the Rakia mission explicitly set out to challenge a perceived failure of the spaceflight industry to sufficiently engage the arts in space exploration. The mission statement asserts that "gravity affects every living being and is relevant for every field of inquiry and all aspects of life," including the arts, and argues that "it seems shortsighted to measure the impact of the space industry simply in missions completed, patents registered or economic value achieved. Rather, it is crucial to see how it improves life: by … offering an expanded canvas for human ingenuity and imagination." The statement identifies that the economic conditions that dictate the nature of current space exploration initiatives mean that space has not yet been fully democratized, but in anticipation of that democratization it is important to sponsor activities and organizations that cannot independently afford to engage with the environment of space; this means "embracing artistic collaborations," as well as sponsoring scientific research (Noble 2022: 10).

One of the Rakia mission's stated goals was to "increase space literacy" (Noble 2022: 21). In order to achieve this, the mission identifies three communities that should be granted access to space: educational, scientific, and artistic. Only when all of these communities are provided with opportunities to engage with the environment of space will space become inclusive, and the full value of space exploration be realized for humankind. Setting an example for other nations and future missions, the Rakia mission employed mission art curators and invited artists to propose artworks for deployment onboard the ISS. The eleven projects that were selected for deployment represent a range of different practices and disciplines, including craft, sculpture, video, augmented reality, and performance, and explored themes including human and non-human dimensions of space and spaceflight. The works included *Impossible Object* (Liat Segal and Yamin Meroz, see Case Study 2) and *Space Jewel* (Shachar Freddy Kislev, see Chapter 3).

Why is space art important?

Udi Edelman, the Rakia mission's art curator, stresses the value of art "as a catalyst for new possibilities and for reflection on what it means to be human" (Noble 2022: 67). The spaceflight industry has elected to send human astronauts into space, and in doing so has acknowledged the value that humans bring to space exploration that is beyond the value of quantitative data that could be generated by robots sent on equivalent missions. So long as the spaceflight industry elects to send humans into space, they must concede that the question of what it means to be human is a significant one. Throughout most of human history, to be human has meant to be an Earthling. Space travel has introduced the possibility of humankind becoming an interplanetary species and has challenged established ideas about human identity.

Humankind distinguishes itself from robots as a social, cultural animal. Advocating the involvement of artists in space exploration initiatives, Pell et al. (2014: 181) write that "artists and cultural professionals are ambassadors for human expression, experimentation, and exploration." The occupation or colonization of any new territory means carrying those social and cultural aspects of humanity with us on the journey. Artist Miha Turšič (2013) refers to a need for the "space culturalization" in which all aspects of humanity are represented in space. Turšič tells us that:

> Space programs deal with … sciences … These sciences are engaged in what is around us but not in *who* are we. We know how old the universe is, how far the galaxies are, what color they are, we know all the planets, yet we do not know what *we* are in space. These programs do not deal with art or humanism. They do not question what it is like to be human in space.

Turšič is co-founder of the Cultural Centre of European Space Technologies (Kulturno središče evropskih vesoljskih tehnologij, KSEVT) in Vitanje, Slovenia, an institution that supports and promotes arts and cultural activities related to space exploration. KSEVT, along with the International Astronautical Federation (IAF), is among several organizations that have cultivated discussions about the inclusion of the arts through broader discussions on the "cultural utilization of space" (Triscott 2012). Such organizations are founded on the belief that human cultural life must be considered in space exploration activities. The International Academy of Astronautics (2009: 12) asserts that the democratization of space must mean the inclusion of diverse minds, including those in the arts and humanities. This must go beyond tokenistic inclusion as an afterthought, driven by "simplistic views of art as merely decorative [and] aesthetic." The Academy argues that the arts and culture must become as integral to spaceflight as science and technology, and "recommends establishing a paradigm for art utilized in human space projects that focuses on art as a core factor enabling mission success, rather than as aesthetic afterthought." Without the inclusion of the arts in spaceflight activities at the core of all their activities, the sector cannot be considered to be inclusive or democratic.

The inclusion of the arts at the core of all space exploration and spaceflight initiatives means including artists as well as the art that they produce, in ways that benefit all members of those communities. Multi-disciplinary communities are capable of diversity of thought, and benefit from the different approaches and methodologies that are employed in the arts. Speaking from her personal experience of working with members of the spaceflight industry, Lisa Pettibone suggests that scientists "are not trained to investigate things [like artists are]." Artists are "trained to be receptive," to consider grand questions, whereas scientists "look in quite siloed detail … [artists] get them to open up and look out." The culture of experimentation in the arts also has a lot to teach the spaceflight industry. The arts are forgiving of errors, to the extent that errors are not viewed as such, but rather as an integral part of research and discovery (vanEenoo 2012). For the arts, errors are interesting and meaningful. One of many ways in which this attitude could benefit the spaceflight industry is that it would offer a second life to artefacts that are considered to have failed or become obsolete by the space sector. Marek Kukula, public astronomer at the Royal Observatory in Greenwich, describes capturing a photograph of a "blank piece of sky," having incorrectly set up the coordinates for the quasars that had been his intended subject. For science, this image might be viewed a failure, but for artist Aleksander Mir (2017: 45), "this has massive value in terms of art … science wouldn't value you staring into the void … [but] art does."

The value of integrating artists into scientific communities has been demonstrated by initiatives throughout the past half-century. The Artist Placement Group was formed in 1966 to place artists in a variety of settings, including government settings, not with the expectation that they would produce artwork, but to cultivate formal encounters between artists and non-artists in ways that would encourage "curiosity-driven work," and to disrupt thinking within organizations in ways that would ultimately make them more adaptable and capable (Rycroft 2019). Today there are similar initiatives throughout the world, and these have been found to result in "renewed enthusiasm and curiosity within [scientists'] research labs" (Lau et al. 2022). One of the most prominent artists' residency programs in operation today is Arts at CERN, which provides opportunities for artists to "engage in dialogue and exchange with the physicists, engineers and staff of the Laboratory." James Wells, CERN theoretical physicist, records benefiting from these residencies by "having someone around who saw the world in a different way, whose influence … could shake up accepted mindsets" (Triscott 2015).

Where government space organizations have failed to take full advantage of opportunities to involve the arts, commercial organizations have stepped in. Most of the spaceworks that have been successfully deployed in space to date have been the product of initiatives from commercial spaceflight entities, including small, specialist organizations that produce tools and equipment for use onboard the ISS. While the ISS is a government facility, the result of cooperation and collaboration between NASA, Roscosmos, ESA, CSA, and JAXA, it allows, and has become partially dependent on, the activities of commercial organizations.[1] This relationship is mutually beneficial, providing funding that supports the ongoing operation of the ISS while creating opportunities for private companies to conduct research in space without the need for their own, private infrastructure. The *Moon Gallery* illustrates the nature of the opportunities for artists to engage with the environment of space at the current stage in the evolution of commercial spaceflight industry.[2] The *Moon Gallery* is a private initiative, funded by the Stichting Moon Gallery Foundation of the Netherlands, and enabled by Nanoracks, a commercial provider of space payload services. The *Moon Gallery* is possible, therefore, largely due to the willingness of governmental space organizations to cooperate with commercial bodies.

Some commercial organizations have recognized the value of the arts to demonstrate the potential of their technologies or services. Many such commercial entities express frustration that governmental organizations are not keeping pace with their ambitions, and have taken it upon themselves to pioneer in areas that have not yet been sufficiently developed by those governments. Many of these are small, specialist companies, such as Silicon Valley start-up, Made In Space, which manufactures 3D printers for space. Co-founder Jason Dunn says, "We're all ... frustrated that the space industry isn't moving as fast as we think it could ... I want to design a future where we thrive in space. Thriving in space is a lot more than just sending the people and the technology to keep them alive; it's about sending the culture and what it means to be human" (Gever 2016). He recognizes that, the primary function of Made In Space printers will be to print functional objects such as tools for astronauts, but considers it important to demonstrate the creative potential of that same technology. In order to test their technology, and to prove its potential for both functional and creative use, Made In Space invited artist Eyal Gever to create a work that would be 3D printed onboard the ISS (see Case Study 5). The resultant work demonstrates that commercial, technical, and creative ambitions can align in the production of artefacts in space.

Barriers to success in art/space collaborations

So long as spaceflight continues to be associated primarily with scientific and technological progress, artists will experience pressure to present their ideas in ways that are accessible to scientific communities. Interactions

[1]This stage in the expansion of the spaceflight industry is essential to Provide opportunities for commercial spaceflight providers to demonstrate their capacity for spaceflight. By providing commercial transportation to and from the ISS, private spaceflights such as SpaceX and Boeing/ Blue Origin have opportunities to test and evidence the effectiveness of their products without the need to first establish their own infrastructure.

[2]Case Studies 3 and 4 feature works that were sent to the ISS in the *Moon Gallery*.

between artists and the spaceflight sector demonstrate that most involved parties still approach microgravity environments as labs, and any activities that take place within them as experiments. Frank Pietronigro's *Drift Painting* (1998, see Figure 2.3) was performed at a time when NASA had a strict policy of allowing only scientific research onboard its parabolic flights. Pietronigro faced opposition from NASA, who, on hearing that his intention was to produce arts rather than science, canceled his flight. After negotiation, Pietronigro was informed that he could go ahead, but only on the condition that he perform science and not art. The artist was instructed to reformulate his plans for the flight, avoiding the terms "paintings" and "performance." Translating his objectives into language that would be more palatable for NASA, Pietronigro proposed observations of "fluid dynamics in microgravity," including observations of "the collision of hard and soft objects" (Pietronigro 2000: 172). He adapted his plans so that the activities would be more akin to a science experiment, mixing a variety of other substances with his paint and introducing ground-based control experiments.

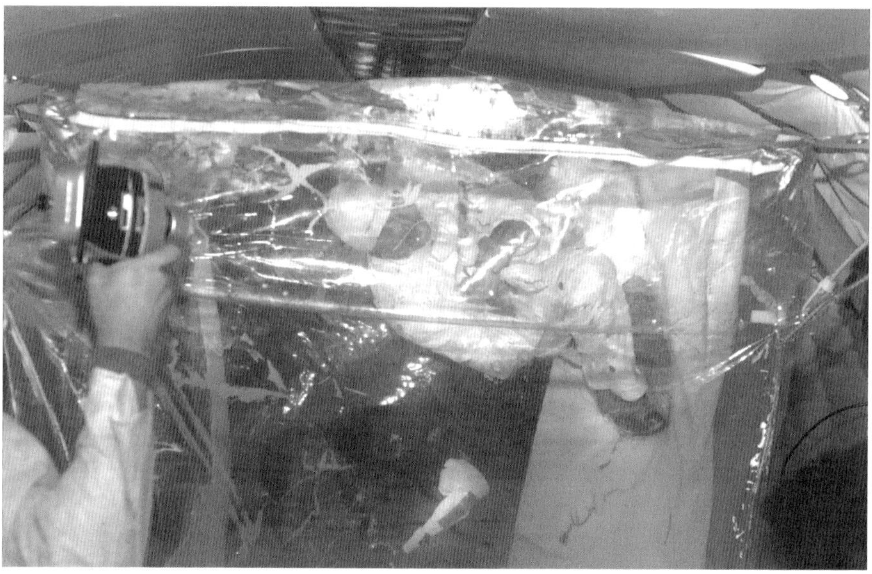

FIGURE 2.3 Frank Pietronigro, *Drift Painting in a Microgravity Environment*, three-dimensional acrylic painting, diamond dust and metal filings, 1998. *Drift Painting* was performed onboard a NASA parabolic flight, at a time when NASA had a policy of only supporting scientific research. In order to be granted permission to go ahead with his performance, it was necessary for Pietronigro to translate the description of his proposed project into terms that would be more meaningful for the scientific community. Pietronigro replaced the terms "painting" and "performance" in this proposal with phrases including "fluid dynamics" and "collision of hard and soft objects." © NASA/Frank Pietronigro

Having adopted language and methodologies that would appease the NASA scientific community, Pietronigro was permitted to proceed.

More recently, artists have reported a shift in culture toward a spacefaring community that is more accepting of "art for art's sake," but, nonetheless, a feeling prevails that microgravity projects should be presented and performed as structured experiments. Alexis Hope found that MIT's Space Research Initiative was open to accepting proposals for projects that were explicitly described as "craft," and yet Hope still felt some pressure to describe and carry out her experiments in structured, "utilitarian" ways (see Case Study 8). Jeanne Morel (see Case Study 7) has similarly observed increasing acceptance of, and enthusiasm for, art in microgravity, but that these views are not yet widespread. In Morell's years of experience onboard ESA parabolic flights, she has found that the ESA community has become more readily accepting of the arts and their different methods, but now finds that the press are responsible for perpetuating a reluctance to accept what art can bring to spaceflight, and persist in describing space art in terms of scientific experiments. In my own experience of presenting research at spaceflight industry conferences, I have found that arts research can be polarizing. While some spaceflight industry audiences are enthusiastic about the potential for space art, others struggle to understand what purpose it may serve, given an absence of quantitative data to demonstrate that space travelers will want to encounter art in space.

Creative engagement with the environment of space is only possible through collaboration. Artists cannot access space without cooperation from spaceflight providers, and some form of collaboration is necessary in the deployment of the works once they are in space. The nature of this collaboration varies from mission to mission. One way in which artists are reliant on cooperation from the spaceflight industries is their dependence on astronauts to deploy their work. When Eduardo Kac (2021: 82) was given the opportunity to propose an artwork for the ISS, it was necessary for him to produce work that was not a pre-formed object, but rather an "art protocol"—a set of step-by-step instructions that could be followed by ESA astronaut Thomas Pesquet, making use only of materials that were already onboard the ISS. Kac worked closely with Pesquet prior to the launch to rehearse the act of production and ensure that *Télescope Intérieur* (2017) would be constructed as per his vision and within the strict timeframe allowed in Pesquet's mission schedule (see Case Study 6). Meroz and Segal similarly describe the requirement to send very explicit instructions and a video recording of how *Impossible Object* should be deployed (see Case Study 2). Like Kac, they had several training sessions with the astronaut before he traveled into space, to ensure that the work would be deployed exactly as the artists intended.

The requirement for step-by-step instructions is routine for astronauts, arising as a result of the need to conduct many complex tasks on behalf of ground-based researchers who are unable to travel to space alongside their own experiments. Astronauts carrying out this work are provided with detailed instructions known as Operations Data Files (ODFs), which exist as scripts or cue-cards, often in portable or wearable formats (Hoppenbrouwers et al. 2017). ODFs are strict, detailed, and do not allow room for error or interpretation. In the case of the deployment of artwork, instructions, equivalent to ODFs, dictate the way that astronauts should deploy the work, and in some cases also how they are expected to interact with and respond to it. When ODFs describe the deployment of an artwork they may be compared to "event scores," a model established by conceptual artist George Brecht and practiced by artists such as Yoko Ono, which could be performed by the artist themselves or by a third party following written instructions (Kotz 2007: 4). There is, in event scores, a "relinquishing of authorial control" by the artist, moving beyond the "single, original expression" to a "repeatable model" (Robinson 2009: 79).

Researchers from other disciplines—including the sciences—describe the frustration of not being able to deploy their work directly. Any experiment that is sent to the ISS or to space onboard commercial sub-orbital flights is deployed in a similar way, by astronauts who have had little or no involvement in its design. Researchers in any discipline have the option of sending their research into space either in an autonomous payload or as "human tended research" (Karika 2023). Where payloads are autonomous, they must be designed to work entirely without human input, generating data within a sealed unit, with the expectation that researchers will retrieve that data once the payload has been returned to Earth. Where research is "human tended," it is often done so by astronauts on behalf of the researcher, following an ODF. Astronauts selected for space are not necessarily skilled in conducting research in the full range of disciplines that are represented in the experiments that are sent to space (Mir 2017: 11). Astrochemist Helen Fraser describes how "scientists make big lists of how to run an experiment on the ISS but often there is a gulf between the astronauts who press our buttons, and us who have designed the experiment" (Mir 2017: 10). The scientific remit of the NASA and other space exploration communities doesn't protect science research from the same frustrations that are faced by artists. The problem, therefore, does not arise solely as a result of a lack of understanding of the arts, but is a more fundamental issue that arises from the fact that there is greater diversity in the projects that are transported to space than there is in the people that accompany them.

This problem can be addressed in several ways, but primarily through recognition that art is only made meaningful through its relationship to humans. JAXA (2016: 8) recognizes an opportunity to promote closer relationships between the sciences and arts in the need for astronauts to deploy work on behalf of artists. This relationship enables astronauts to become co-producers of artwork, and this opportunity may be embraced by some artists who value the unexpected outcomes of stakeholder interaction. However, such interactions only become co-production when the astronaut is able to make their own independent decisions about how they will respond to the work. The democratization of space will eventually lead to a community of astronauts with a more diverse skillset, and eventually artists will be able to accompany their work to space so that they can deploy it themselves. In the meantime, greater collaboration between the spaceflight industry and astronauts on Earth, prior to the deployment of spaceworks in space, has the potential to inform the way that research and art is deployed, as well as the ways that astronauts approach that deployment. Allowances need to be made for astronauts to take time to respond intuitively and in their own time to the artworks that they deploy.

Though there are many benefits to collaboration between artists and scientists in the spaceflight industry, these collaborations are not always inclusive, and can perpetuate several myths about art/science collaboration: first, the myth that the arts have more to learn from the sciences than vice versa; second, that projects for space must necessarily incorporate advanced technology or exploit scientific innovation or that artists' engagements with the environment of space would be inadequate unless guided by representatives from the sciences; and third, that art and science are two distinct disciplines. The notion that art and science are disciplines that exist at opposite ends of a spectrum is relatively new (Efland 1988; Wilson 2002: 5). The reality of arts practice is that it is inherently interdisciplinary, and artists are accustomed to engaging in practice that challenges disciplinary boundaries. Art and design "are in a continuous state of evolution and cross-fertilization" (QAA 2019: 3). Experiments with technology and approaches that challenge disciplinary boundaries, including collaboration, have defined the field as one that is in a constant state of evolution, with pregnable and ever-changing borders. Collaboration exposes artists to different ways of thinking and working that enable them to take more critical approaches to their own practice, and to consider their work in new contexts.

While artists may often transgress disciplinary borders, it is sometimes in ways that are not recognized, because their approaches to gathering and deploying knowledge may be different to those that are associated with those other disciplines. For example, artists typically do not have a scientific or mechanical knowledge of gravitation, and yet they routinely produce work that is informed by gravity and its effects, for example, by experimenting with weight and balance (see Chapter 3). These same artists, who gain an understanding of the likely behavior of their materials through "observation" and "speculation," are assessed (first during their education, and then later by others in the arts community) on their ability to "experiment with ... materials, processes ... and environments" (QAA 2019: 7; 13). The education of artists encourages them not to take for granted the physical conditions in which they operate, and to question and explore the limits of the tools and environments with which they are working. Arguably, therefore, strange new worlds are native territory to the artist. When faced with new environments, such as the environment of space, artists' instinct is to apply that same approach to the acquisition of knowledge that they have employed throughout their careers, that is, to learn through experience and informal observation. Faced with the need to learn about microgravity, for example, direct experience and observation of microgravity's effects would be more in keeping with the way that artists learn than collaboration with scientists who pursue a mechanical or mathematical understanding of microgravity.

"Experiential knowledge," observe Niedderer and Reilly (2010), "is not often associated with research and organized inquiry, and even less often with the rigor of debating and honing research methods and methodology." Catherine James (2018: 3) likens artists' experiments with gravity to "the simple empiricism of early experiments in physics," that were essential in Newton's early revelation about gravity, but that have since been superseded in the scientific community. Differences and divisions in approaches to research and knowledge production are, argues Ariane Koek (2017: 3), a barrier to effective collaboration between the arts and sciences. There is a perception that the arts are not equipped to deal with the environment of space, and that they must adapt their methodologies, or make some concession to the sciences, in order to produce artefacts for this environment. At the same time, there is a perception that science must be dumbed-down in collaborations with artists; that the sciences "operate at scales and levels of complexity" that cannot be understood by those outside of their field (Crisp and Triscott 2017: 11). These perspectives are problematic for two reasons: first that they do not conceive of different methods of knowledge production as equally valid, and, second, that they presume that hypothetical, scientific, or mechanical understanding is more valuable than experiential knowledge.

To some extent, this problem arises from the different ways that artists and scientists describe their research. The ETTAS found that "many commonly used terms in astronautics do not translate directly to the frames of reference in the arts and vice versa," including the term "research," which was differently understood by scientists and artists in the team (Pell et al. 2014: 182). This problem was multiplied in conversations observed by the ETTAS because collaborators had not initially recognized that their understanding of the term was different. Terminology was "unwittingly exchanged without clarification" in these conversations, creating misunderstanding and barriers to progress. Advocating for the use of a mediator or translator in art/sci collaborations, Koek (2017) observes that the arts and sciences use very different, sometimes incompatible languages and ways of looking at the world. These languages and approaches can alienate artists, force concessions, and compromise the integrity of resulting artefacts.

If artists learn most effectively through experience, then the nature of collaborations with the spaceflight industry may need to be reconsidered. Partnering artists with scientists may be less valuable than providing artists with opportunities to acquire experiential knowledge through intense programs of observation and experience, in which they are able to witness and sense the environment of space and its effects, or to engage in exploratory conversation with astronauts and others who have first-hand experience of that environment. Access to spaceflight experience is increasingly available as parabolic and sub-orbital flights become more accessible. Increasingly, participants do not require specialist scientific or technical expertise to take part in these flights. Experiential knowledge can also be gained through indirect observation, as by, for example, viewing video recordings of the effects of weightlessness that are publicly available via NASA. With access to opportunities to acquire experiential knowledge, artists are able to independently develop the knowledge that they need in order to produce work for the environment of space.

Where collaborations between artists and scientists do not yield appropriately innovative outcomes, it may be in part due to the kind of artists who are invited to take part in these collaborations. Artists make the case that it is not only important to acknowledge the value of the arts, but also to include the right kind of artists in art/space initiatives. Nicola Tricott (20102) observes a lack of equivalence in the selection criteria that government space organizations employ for the arts and sciences, setting the bar for inclusion much higher for scientists than for artists. While government space organizations invite contributions from "cutting edge, peer-reviewed scientific research," those same organizations' art initiatives "have tended to come about through personal interests and contacts of individual space agency personnel," particularly when art is viewed as an outreach activity, "rather than through an institutionally-recognized professional engagement with art experts." A lack of knowledge about how to recognize excellence in the arts means that there is a risk that the art produced as a result of these initiatives is of limited value, and does not live up to the potential of arts that take place at the forefront of practice. The consequence of a failure to engage appropriately with the right members of the arts community is potentially significant, not only for artists but also for the whole spacefaring community and all its stakeholders on Earth. At best, argues Lisa Pettibone (2018), "[professional artists] won't get to decide what art goes into space and we could end up with a lot of bad art in space." At worst, a growing human population in space will be limited in its culture and achievements because it does not represent the diversity of human thought that exists on Earth.

The inclusion of qualified artists, recognized as practicing at the forefront of their discipline, may help spaceflight organizations to address the unidirectional knowledge-exchange that has characterized many of their past initiatives. Years of engaging with hobbyists as opposed to professional artists has led to a siloing of the arts, that appears to manifest as a lack of respect or reluctance to include artists at the forefront. Artists interviewed for this book stress that there is not necessarily explicit resistance from the spaceflight industry to engaging with the arts, rather, there is a lack of understanding about art and the value that it may add to space exploration initiatives. Even where artwork has been supported by the spaceflight community, the artists who produced that work are often kept at arm's length. Lisa Pettibone (2018) describes a situation in which there is "not enough communication on both sides," leading to missed opportunities for artists and scientists to identify common interests or to discover the valuable differences in their approaches to problem-solving. In order to fully benefit from the inclusion of the arts, the spaceflight industry needs to consider the value of art as well as artists, whose different approaches to enquiry have much to contribute to the diversity of thought that needs to take place for humans to thrive in any new environment.

Case Study 3
Max Baraitser Smith,
Lunar Bureau of Weights and Measures, 2021

One cubic centimeter is the exact size of the sand-filled acrylic cube that was Max Baraitser Smith's contribution to the *Moon Gallery*. Like all of the other works in the *Moon Gallery*, Baraitser Smith's work was required to fit inside a 1 cm^3 box (see Figure 2.5), but for Baraitser Smith this measurement has additional significance. The cube is intended as the first prototype measure for a new lunar system of measurement, and the founding object of the Lunar Bureau of Weights and Measures, a hypothetical organization that Baraitser Smith proposes as the lunar equivalent of Earth's existing Bureau of Weights and Measures, the organization responsible for establishing and maintaining uniformity of units of measurement on Earth.

There are numerous artificial objects on the Moon that were designed using metric units, but there is no lunar standard on which those units are based. Baraitser Smith argues that the use of metric units on other planets is problematic, due to their terrestrial-centric history. The length of a meter was initially defined as "one ten-millionth part of the earth's quadrant" (US Department of Commerce 1932: 3). Thus, the meter, and the metric system more broadly, is a system of measurement that is Earthly in its origins. The metric system is not only a product of the Earth, but also of human culture, arising from the peculiarly human habit of inventing and imposing standards as a way of quantifying and understanding the world, and standardizing products and processes through the application of those arbitrary systems. The establishment of a lunar standard of measures would be one way in which humankind could establish this aspect of human culture on the Moon. Baraitser Smith's proposal to establish a lunar standard of measurements therefore both extends human culture into space and simultaneously encourages a transformation of that culture as it becomes established on other planets, in a convergence of human and extra-terrestrial cultures that becomes native to humankind's colonized planets.

Just as the meter was initially defined in relation to the circumference of the Earth, Baraitser Smith's lunar measurement, the new lunar *lune* (abbreviated to ln), is defined in relation to the circumference of the Moon. The lune is defined using the same method as the Earth meter, as "one ten-millionth of the distance between the North Pole and the equator … this time based on the moon's circumference" a measure that is, ironically, measured in Earth meters.[3] Baraitser Smith's acrylic cube is a symbolic

[3]The history of artworks proposed for the Moon highlights the arbitrariness of standardized measurements. The *Moon Gallery* is based on Earth's metric system, with its works displayed in a "10 cm × 10 cm grid structure." Alexander Zaklynsky's (2018) initial proposal for the *Moon Gallery* identifies foundations for the project in the 1969 *Moon Museum*, which was measured in inches, and described by Zaklynsky as "three-quarters of an inch by half an inch in size."

object that represents the relationship between the Earth meter and new lunar lune. The 1 cm^3 cube contains exactly 1 cubic *centilune* of sand, 419 grains collected one-by-one, using tweezers, from a beach in Scheveningen in the Netherlands. By containing both measures, the cube places these two measurements in relation to one another. The object may also have the practical purpose of serving as a conversion tool, while the volume of sand can act "as a benchmark for manufacturing new measurement equipment: rulers, tape measures, beakers, pipets."

Baraitser Smith views the project as having three phases, each aligned to a different location: the first phase was the collection of the sand and production of the acrylic cube on Earth; the second was the launch and journey through space; and the third is the siting of the work on the Moon. Baraitser Smith describes a kind of absurd, performative activity of sand collection, in which he painstakingly plucked each grain of sand from the beach, then assessed each pinch to decide whether he felt it constituted a useable grain or unusable dust. In Baraitser Smith's original proposal, this sand would be loosely enclosed in the cube so that it would be activated by weightlessness during the journey (see Figure 2.4), then settle into a new configuration once sited on the Moon. In practice, safety restrictions meant that the sand had to be set in solid acrylic due to concerns about grains escaping into the cabin of the ISS.

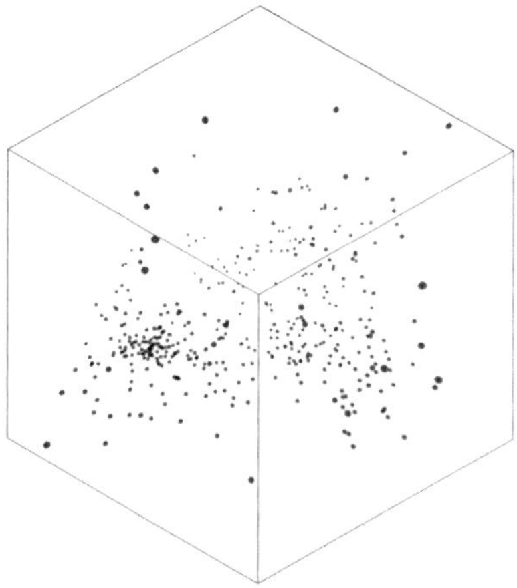

FIGURE 2.4 Max Baraitser Smith, *Lunar Bureau of Weights and Measures*, 2021. Visualization of a cubic centimeter of acrylic, containing a cubic centilune of sand, depicted as if floating weightlessly. Baraitser Smith's proposal described grains free to reconfigure within the cube while onboard the ISS, and then settling into a new form once the cube had been deposited on the surface of the Moon. In practice, ISS safety concerns meant that the sand had to be set in a fixed arrangement in solid acrylic. © Max Baraitser Smith

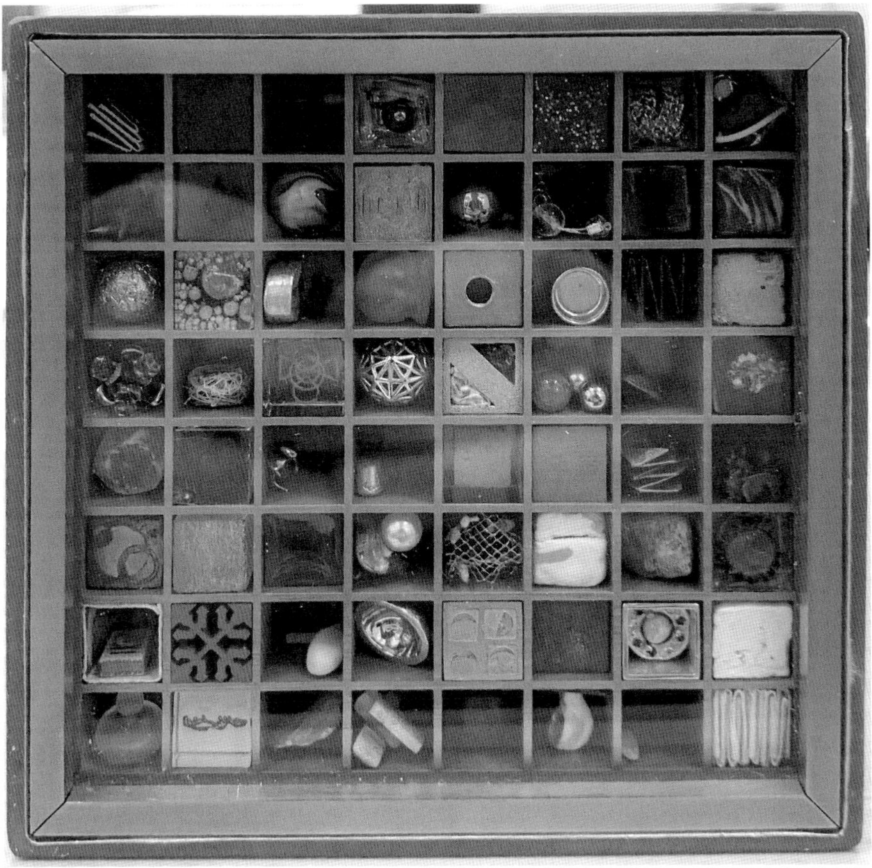

FIGURE 2.5 *Moon Gallery*, containing sixty-four artworks, each housed in a 1 cm^3 box. *Moon Gallery* is a private initiative, funded by the Stichting Moon Gallery Foundation of the Netherlands, and enabled by Nanoracks, a commercial provider of space payload services. Works in the *Moon Gallery* include three featured in this book: Max Baraitser Smith's *Lunar Bureau of Weights and Measures*, Lisa Pettibone's verdant (see Case Study 4) and Arti Struyanskiy's *Human Frontiers* (see Chapter 4). Image © *NASA*

In the context of this book, Baraitser Smith's unplanned plan for free flowing sand is a significant part of the project. His proposal imagined that the 419 grains of sand would "create new patterns and sculptural forms" within the microgravity environment of the ISS, or onboard the rocket that would eventually transport the *Moon Gallery* to the Moon. Weightlessness would cause the sand to "become a free-floating cloud of individual grains, occasionally colliding," and each collision would create a "tiny sonic event." Baraitser Smith describes the differences between the dune-like form that the sand would adopt when it settles on the Moon, and the separately floating, weightless grains, thus distinguishing between the Gestalt whole of the sand when held together by gravity and the separate particles that gain their independence in microgravity. While weightless, the sand would be activated, coming to life as hundreds of distinct objects, each with its own destiny. Then, on the Moon, they would resettle and once again be constrained by gravity into a whole.

Baraitser Smith's use of sand, and his interest in space, was inspired by his undergraduate experiments with the acoustic resonance of metal plates. In these experiments, Baraitser Smith dropped grains of sand and salt onto plates and explored techniques that would extend the length of the ringing sound that is generated by this activity. The project yielded the realization that air resistance was responsible for preventing the resonance from continuing forever, and that therefore, hypothetically, if the experiment could be reproduced in space, he could create an instrument that "would ring [and ring] and never stop." Sand, which already has associations with the passage of time as a result of its use in hourglasses, became, for Max, associated with the potential for transcending space and time. In an hourglass, sand shows us that a single unit of time is finite; but in microgravity, where sand cannot fall, it cannot show the passage of time, and we are reminded that time is infinite.

Art and science unite in Baraitser Smith's consideration of sand, time, and resonance: he practices at the intersection of art and science. He studied at the ArtScience Interfaculty in the Hague, and feels it would be inappropriate to make a distinction between his dual roles as an artist and as a scientist. He approaches this distinction on a project-by-project basis, classifying his *Moon Gallery* project as art. If the project were a science experiment, Baraitser Smith would have expected to conduct trials into the sand's behavior and build a model that would predict the ways that it might move in microgravity. When producing art, Baraitser Smith is free to embed unknowns into the work, and in this case to imagine endless possible configurations for the weightless sand without declaring that one might be preferable, or more likely than others. Allowing room for "possibilities" and "imagination" also means that it doesn't matter that the "tiny sonic events" cannot be heard by the human ear, just as Baraitser Smith's hypothetical outer space resonating plates could not be heard without atmosphere to carry the sound. These ideas exist within the "imaginary speculative realms," not as proposals for actual artworks, but rather as "thought experiments," like those conducted in the sciences.

Approaching this project as an artist has allowed Baraitser Smith to embed political significance in the work, and to produce an outcome that would embody play and unpredictability. As art, the work is able to function as a critique of the arbitrariness of systems of measurement, and of humankind's absurd desire to standardize and systemize our world. In short, Baraitser Smith, identifies the work as "a kind of joke." Baraitser

Smith wrestled with the question of whether this spacework could be presented as satire. He recalls that the main challenge of his developing proposal was finding "the line between a real proposal and a joke," and even considered the possibility of producing a parody of the logo of the Bureau of Weights and Measures. He concluded that to present the work as an outright parody would undermine the serious, political, and ethical significance of humankind's colonization of other planets. Yet, at the same time, the work could not be overtly political. At this current period in the history of space art, when artists are still fighting for recognition of the value of their contribution to the sector, it is too early to propose work that is overtly political or disruptive.

The significance of the idea, more than the significance of the object itself, means that Baraitser Smith is undeterred by the knowledge that the work will have "no predefined audience." Existing in space, where only a few astronauts will encounter it, the physical manifestation of the project will be seen by very few humans. The idea, however, suggests Baraitser Smith, can reach audiences who do not directly encounter the cube. He has considered developing an online presence for the Lunar Bureau of Weights and Measures to grant larger audiences access to the idea. This website would not serve the purpose of displaying the work, but rather of provoking questions about ownership and representation, "especially art objects and their role" in representing humanity in the colonization of other planets. "Inter-global and inter-nation structures" argues Baraitser Smith, need to be interrogated, along with the place of the arts and in relation to those systems and structures, "because those structures are usually not representative."

More broadly significant, suggests Baraitser Smith, is the fact that this and other artworks have been sent into space. Space is the most inaccessible location known to humankind, and is even more inaccessible to artists than to scientists; the opportunity to "play on a spaceship" is more expensive than any other opportunity for play that money can buy, "but because of that it's probably the most important." Spacefaring activities will define the future of humanity, and it is important that the arts demonstrate that they are equipped to engage with unknown, inaccessible frontiers if the arts sector is to be included in humanity's future.

Max Baraitser Smith was interviewed on October 11, 2023.

3

Post-gravity Thinking: How Weightlessness Changes Everything

"Entirely different foundations": *Normogravity* is not normal in the universe

In order to consider the ways that artists may engage with the environment of space, it is necessary to identify the features that differentiate that environment from those that artists encounter on Earth. Spaceworks may be located in open space (the extravehicular environment), as in art satellites, or aboard a spacecraft or space station (the intravehicular environment). These environments present different challenges and conditions to artists. Most human spaceflight activity takes place in the intravehicular environment, and it is this that is the focus of this chapter. Aboard a spacecraft, the atmosphere is controlled, and humans and their cargo are protected from the vacuum of space. This environment differs from an atmosphere-controlled environment on Earth by its gravity conditions. While the atmosphere onboard a spacecraft may feel very similar to that onboard an aircraft in many other aspects, the weightlessness of the crew and surrounding objects makes the experience of spaceflight markedly different from that of air travel. Weightlessness causes objects and materials to behave differently, and enables encounters that would not be possible on Earth's surface. For artist Eduardo Kac (2005: 22) it is not space, broadly, but more specifically, "zero gravity [that] is the next frontier." Unlike other features of the environment of space, there is a "unique specificity" to weightlessness that "cannot be reduced to other conditions" (Kac interview 2023). Kac observes that "everything that we do, without exception, is effected by gravity," and he argues that, therefore, all human-made forms may become "radically different" when intended for deployment in weightlessness. Maja Murnik (2016: 68), reflecting on the similarities between the virtual and microgravity environments, suggests that "art beyond gravity derives from entirely different foundations" to terrestrial art practice.

In the arts, the term *weightless* is often used with hyperbolic, metaphoric, or poetic intent (Brownie 2019: 42–3). In the art/sci context, it is necessary to pursue a common understanding of the term, in order to enable clear communication between artists and space scientists, and to provide artists with an understanding of the conditions that shape spaceworks. The arts and science communities have specialist notions of *weight* that differ from, or are more precise than, the layperson's understanding of this term. Taibu et al. (2015: 1) observe difficulties and inconsistencies in explanations of *weight* and *weightlessness* for lay audiences, identifying "both conceptual and language difficulties" that are encountered when proposing definitions of these terms. The force that an object exerts on a set of scales varies in different gravitational conditions, and so it is necessary to disentangle weight from its measurement. The field of physics offers a "gravitational definition of weight"—the gravitational force on an object—and an "operational definition of weight"—the force between an object and a measuring scale (ibid., 2). Both of these can be distinguished from *mass*, which is more akin to the artist's or designer's understanding of the term.

In the visual arts, the term *weight* is typically used to describe "visual weight" or "compositional weight." In two-dimensional composition, weight is suggested by size, shape, color, and location within the frame as well as location relative to other objects within the composition (Rudolf 1974: 23–6; Donis 1973: 30–2). In the visual arts, therefore, the weight of an object can be apparently altered by relocating or changing the color of that object, and these factors may counteract each other (Rudolf 1974: 28). Light also plays an important role in the perception of weight. Ken Koenderink et al. (2017: 299) find that audiences often "judge weight based on luminescence." In three-dimensional design and sculpture, both scientific and artistic definitions of weight play a role. While visual weight impacts the appearance of a sculptural object, physical weight may inform artists' choice of materials and other practical concerns about the ways that the work is displayed.

Artists' practice is informed by their experiential knowledge of Earth gravity and its effects. Although not often explicitly acknowledged as such, an understanding of gravity's effects is manifested in artists' consideration of weight and balance in their practice. Wherever sculptors explore the weight of their materials, suspend or balance objects on one another, or simply rest an object on a plinth, they are engaging with the effects of gravity. Similarly, artists who choose materials knowing that they will drape, pool, or rest in a particular way, are employing knowledge of the effects of gravity. Catherine James (2018) observes that artists have "employed gravity as an immaterial ready-made," either consciously or unconsciously engaging with gravity through exploration of balance, weight, and physical resistance. An apparently innate understanding of the effects of gravity is taken for granted as general knowledge, not scientific knowledge, because it is not explicitly taught or acknowledged. For artists and designers, "experiential knowledge or tacit knowledge [is] an integral part of their practice" (Niedderer and Reilly 2010). Much of the knowledge that informs creative practice is not formally recorded or recognized, but is implicitly understood. The science of gravity is not explicitly taught in art schools, and yet artists do learn of its effects through their acquisition of practical and observational skills. It is within this framework that artists "intuit gravity" (James 2018: 3). Familiarity with the effects of gravity, gained through a lifetime of operation within Earth's gravity conditions, is one of the fundamental, if unspoken, principles on which the visual arts and design are founded. Art scientist Max Baraitser Smith (see Case Study 3) illustrates this for lay audiences using the example of riding a bicycle. When we ride a bike, he observes, many of us are unaware of the science that is keeping it upright and preventing us from falling, and yet we rarely fall. When learning to ride a bike we do not learn the laws of physics that keep us upright, and when these laws are taught in school it is not in the context of cycling, and yet we are able to learn not to fall off our bicycles.

Where artists make explicit reference to the effects of gravity, it is often through exploration of behaviors that their materials demonstrate in the Earth gravity environment. Slovakian sculptor Mária Bartuszová (1936–96), for example, "used the gravitational pull on the weight of poured plaster to shape" her work, referring to this process as "gravistimulated shaping" (Tate Modern 2022). Her work includes an array of plaster forms, shaped by their own weight and the weight of other objects that rest on top of them. In one series of untitled works, plaster-filled balloons, alternating with rocks and other heavy, solid objects are stacked into towers, the plaster then left to dry and balloons removed, leaving solid forms that are squashed beneath the objects above them. In *Untitled* (1972–74), four plaster-filled balloons ooze and bulge in the spaces between five blocks of wood. Many of Bartuszová's works are held together only by gravitational forces, so that the parts may be separated and handled as independent forms. Separated from the objects that have shaped them, the echoes of those objects are recorded in their form, a permanent consequence of the force of gravity and the role it plays in shaping forms.

Bartuszová described some of the forms in her early works as "a drop" (*sic*), making reference to the event of falling – the result of an act of dropping – as well as the shapes produced during that process (Tate Modern 2022). *Untitled (drop)* (1963–64, see Figure 3.1), is a large droplet-like plaster form suspended from thread. Bartuszová created this and similar forms by filling balloons with wet plaster, then letting that plaster cool as the balloon was suspended in mid-air (Redzisz 2015). The form swells at its bottom, as if in pursuit of the spherical form that it might eventually achieve if left to fall, while its top is stretched upward into a point, where the attached thread

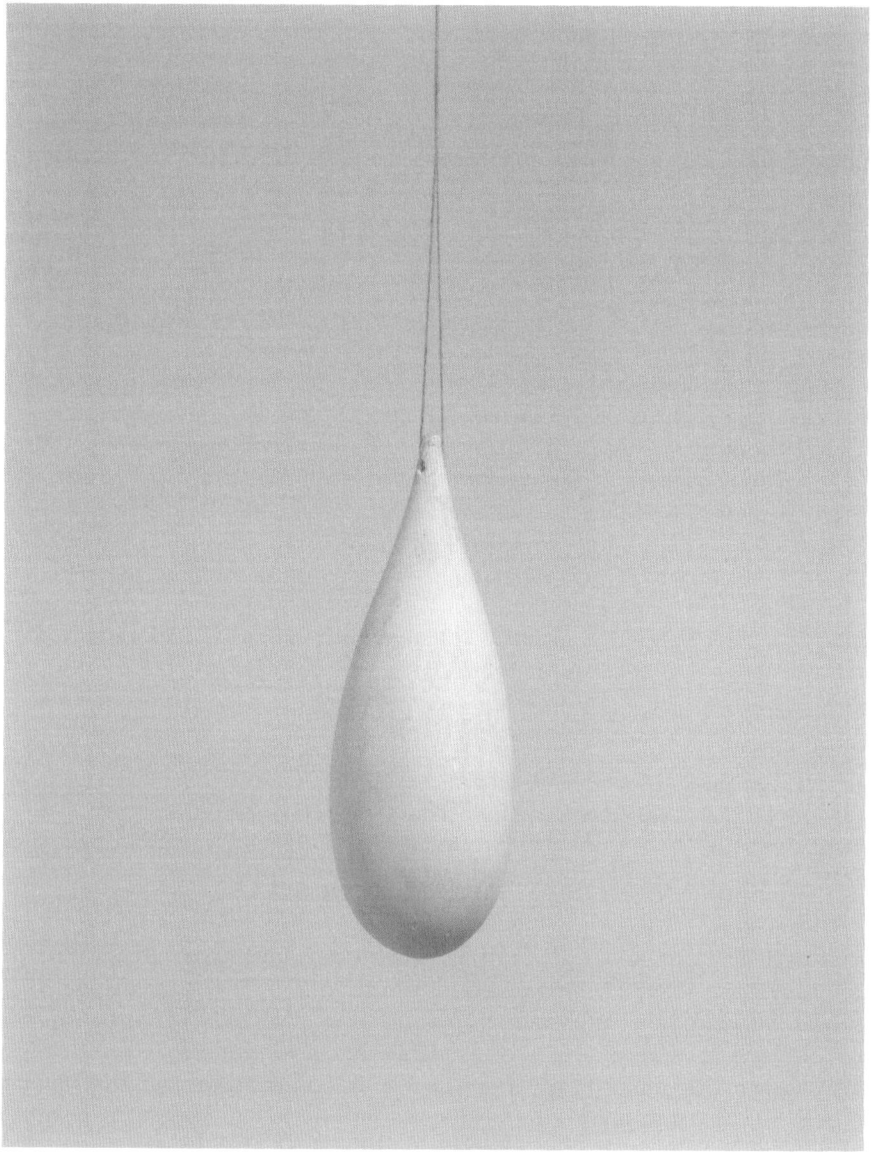

FIGURE 3.1 Mária Bartuszová, *Untitled (Drop)*, plaster, 430 × 160 × 150 mm, 1963–64. Many of Bartuszová's works exploited gravity to shape wet plaster, a process described as "gravistimulated shaping." © The Archive of Mária Bartuszová, Košice

prevents it from falling. On Earth, we are familiar with this shape. Liquid, dropped, is shaped as it falls; it transforms in mid-air into a shape that we understand as a "drop." The act of *dropping*, and the reshaping of liquid or elastic forms during that process, are familiar in Earth gravity, but are not experienced in the same way in microgravity (see Chapter 6). On Earth, we can equate dropping with falling. In space, specifically in orbit, falling is not experienced as a drop, but rather as weightlessness.

Weightlessness is sometimes wrongly equated to zero-gravity or "zero-G" (Karmali and Shelhamer 2008: 595). Weightlessness is the apparent condition of objects in a *microgravity* environment, where the pull of gravity is barely discernible, but never "zero." In reality, the pull of gravity is always present, but not always experienced as such. The term, microgravity, is used in space science in recognition of the fact that all objects with mass have gravity, and so gravitational forces are inescapable. The image of Newton's apple falling toward the Earth gives us an over-simplified impression of how gravity works. Gravity is the physical force of attraction between objects, with all objects exerting force on one another. Earth gravity is not simply a unidirectional force that pulls all objects toward the Earth's core; rather, all objects exert this pull on each other to a greater of lesser degree, depending on their mass. Just as we are pulled toward the Earth, the Earth is pulled up to meet us. Our experience of Earth gravity can therefore be considered a relationship between ourselves and the Earth, as we share a mutual attraction.

What humans experience as the downward pull of gravity is "a duel between the centripetal forces pulling us towards the Earth and the centrifugal ones working to throw us off" (Joseph Lanza, cited in Eshun 2005: 28). The view of the Earth from space, rotating at 460 meters per second, reveals that any stillness that we experience on Earth's surface is an illusion. Logan Smith (2022) observes that our Earth-bound history has created "an intuition that is stillness" that is disrupted in spaceflight. Smith's observations of astronauts aboard the ISS reveals that they "are constantly drifting and touching things to maintain their position relative to something. It's all about relative position." These conditions have been normalized for humans and other terrestrial inhabitants as a result of our being native to Earth, to the extent that Earth gravity is described as *normogravity* or *1G*, and is the standard by which other gravity conditions are measured.

Astronauts in orbit experience weightlessness, not because they are unaffected by gravity, but rather because they are in freefall. Objects in orbit are constantly falling to the Earth, but as they fall, the Earth rotates away from them. They are, therefore, forever falling without ever becoming closer to the Earth's surface. When an astronaut is aboard an orbiting station, they and the objects around them all fall at the same rate, as when a box full of objects is dropped from a height.[1] As the astronaut and their surroundings are all falling together, the astronaut does not experience orbit as falling, but rather as a state of being suspended alongside objects that are also suspended in the same space. An object that is in freefall is not free-of-weight; rather, it is free from opportunities to exert force on a measuring scale. It may be helpful to imagine, for example, an astronaut cupping a ball in their hands. The astronaut will not experience that ball as having any weight, because that ball is not pressing down on their palms. This does not mean that the properties of the ball have changed, but that the astronaut is being pulled toward the Earth at the same rate as the ball. Weightlessness is, therefore, relative and subjective. Taibu et al. (2015: 3) describe weightlessness as a "sensation," as if to highlight the potential subjectivity of this experience of weightlessness.

The same weightlessness that is experienced in an orbiting space station is replicable within the Earth's atmosphere onboard aircraft. So-called *zero-G* aircraft fly in a series of parabolic arcs so that the force of gravity increases to up to 1.8 times its usual force on the upward slope, and then decreases to 0G as the craft pulls into a downward slope on the other side of the arc and enters freefall. During each parabola, passengers experience 20–30 seconds of weightlessness. Parabolic flights offer commercial tourists, researchers and creative practitioners access to a space-like environment in which to experiment, without leaving the Earth's atmosphere. Parabolic flights have been used to train astronauts since the 1950s, and have more recently been available to commercial passengers and researchers through providers such as Zero-G in the USA, Novespace in Europe, and ASTRAX in Japan. The microgravity environment onboard parabolic flights has been used for artistic engagements with weightlessness including those enabled by The Arts Catalyst and MIT that will be explored later in this chapter. The experience of artists Dragan Živadinov, Dunja Zupančič, and Miha Turšič aboard parabolic flights has led to them to identify a new field of practice that they have termed "postgravity art": art that responds to, or engages with, microgravity.

Many other artists and philosophers who have experienced weightlessness also describe what might be termed a *postgravity experience*. This may be characterized as an appreciation that normogravity is not normal in the universe. Normogravity, as experienced on Earth, is the condition that has shaped most human experience throughout our history of existence on the surface of the planet. Gravity does not just rule, but "defines us and builds us in the Earth's environment" and, as a result, has the potential to limit our understanding of what is possible (Murnik 2016: 68). Those who have experienced weightlessness describe re-experiencing gravity as a "peculiarity,"

[1]Tall drop chambers such as the European Space Agency's ZARM tower in Bremen, Germany, enable weightlessness testing on a small scale, inside containers that are dropped from a great height.

acquiring an appreciation for the extent to which their past practice has taken gravity for granted (Eshun 2005: 29). Rob La Frenais (2005: 10) describes his first experiences of weightlessness aboard a parabolic flight as a wake-up call, a moment "where you feel like you have been dreaming for the rest of your life" (*sic*). Philosopher Mikhail Ryklin (2005: 15), reports that "even a temporary stay in states of weightlessness ... changes one's relation to the body and its terrestrial possibilities revealing that "your usual body weight [is] simply one of the possibilities of your body." These accounts suggest that the experience of microgravity contributes to the possibility of a post-Earth, and, therefore, post-human identity, in which the bodily experience is substantially affected by different gravity conditions.

Postgravity approaches to art practice are founded on an understanding of the relativism of gravity. They may explore the potential of microgravity in ways that highlight differences from the way familiar objects behave on Earth, or they may reconsider objects and materials affected by gravity from the perspective of an appreciation that different gravity environments are possible, from a position of understanding that the resulting work would behave differently in other gravity conditions. Postgravity thinking may be demonstrated on Earth, in work grounded by Earth gravity, as much as it may occur in space in a microgravity environment. Those artists who have experienced microgravity find that their practice on Earth is affected by that experience, so that their terrestrial work is informed by a similarly reflective and critical approach to gravity conditions to work that may have produced for or in microgravity. A postgravity work reflects on the extent to which past practice has been unconsciously shaped by Earth gravity and expresses an understanding of the potential to exist in or produce work in other gravity conditions, mainly microgravity.

The arts in microgravity

There is a feeling of exclusivity attached to artefacts that have been produced or deployed in the microgravity environment. Objects that have been exposed to this exclusive environment can be fetishized and commodified (Watlington 2023). As a result, there is a risk that engagements with microgravity can become tokenistic, particularly when performed for the purpose of showing or selling the work subsequently on Earth. Banker-turned artist Nasser Azam's *Homage to Francis Bacon: Triptych I* and *II* (2008), partly produced onboard a parabolic flight in microgravity, sold for a sum of such significance that it funded his transition from banker to professional artist (Usborne 2009). Azam's paintings were begun on the ground, then overworked with oil pastel during a Russian parabolic flight. The microgravity-performed additions to the work are minimal, and the paintings are not significantly transformed by them, but they are discernible in the final work as strokes that are less controlled than, and in a different medium from, the painting beneath. His decision to work with pastel instead of paint minimized the effects of microgravity that have been embraced in other microgravity painting projects such as Frank Pietronigro's *Drift* (see Chapter 2), and maximized the control that Azam would have over his composition. Indeed, the pastel strokes in the completed tryptic do appear far more controlled and purposeful than many of the more recent gestures performed by artists working in paint or craft materials. These effects are achieved despite microgravity, not resulting from it, illustrating efforts to overcome the effects of microgravity rather than to embrace them. Azam credits his background in finance as a reason for his success in an increasingly commercialized art world, informing the way in which he describes his work to potential buyers (Collett-White 2009). The value of *Homage to Francis Bacon* lies largely within the knowledge of its physical journey, and Azam's decision to market the work as "Space Art." It is tempting to write off Azam's microgravity activities as gimmicky, but he has described a profound impact on subsequent work. His time spent in microgravity has "informed new developments" in Azam's practice, intimating that the experience of weightlessness has prompted a long-term shift in his creative and intellectual relationship with gravity.

While postgravity thinking may transform our engagements with gravity and human identity both intellectually and philosophically, weightlessness also presents a material challenge (Kac 2005: 25). Weightless bodies do not perform in the same way as bodies affected by Earth gravity: weightlessness affects the way that objects are oriented in their surroundings (see Chapter 4); the way that human bodies move and function (see Chapter 5); the way that living organisms grow; the way that materials behave (see Chapter 6). The physical effects of microgravity mean that postgravity art has substantially different concerns from those of Earth-bound art, and that this new field of practice will be tangibly different from any material practice conducted throughout the history of art. In this new paradigm, stability is replaced by instability, orientation by disorientation, and absoluteness with relativity (Eshun 2005: 29–30).

Rob La Frenais (2005: 9) locates microgravity art alongside a history of art practice that engages with "experiences that alter the effects of gravity" in different ways such as skydiving, diving and levitation. Kac (2005: 19) cites, among several examples, photographs of László Moholy-Nagy "levitating a chisel with compressed air" (1947), magnetic levitation on Takis's *Télésculpture* (1959) and Hans Haacke's *Sphere in Oblique Air Jet* (1967). Haacke explored "air as a medium," using jets of air to suspend sculptural objects above the ground (Sutton 2019: 11). In *Wide White Flow* (1967) and *Blue Sail* (1964–65), Haacke employed the irregular air flow from electric fans to create fluctuation motion in loose fabric, and in *Sphere in Oblique Air Jet* (1967), a steady stream of air to stably suspend a spherical balloon in mid-air. Such works established the possibility for sculptural objects that do not rest on a surface, and for empty space to contribute materially to a work. In microgravity, suspension in the void is a natural state, and weightlessness can be employed as a medium. Given the history of art that engages with suspension in mid-air, it is inevitable, suggests Kac, that artists will seek opportunities to produce works that "engages outer space materially" through experiments with microgravity.

Thanks to initiatives by The Arts Catalyst, MIT Media Lab, and l'Observatoire de l'Espace at CNES, parabolic flights have been the site for creative engagements with microgravity by artists, designers, and performers, all of whom have contributed to the emerging field of postgravity art. The UK-based Arts Catalyst supported parabolic flights between 2000 and 2004, flying from Star City, Russia, and Bordeaux, France. Participants in these flights included artists, philosophers, choreographers, and performers, as well as reporters and scientists (Triscott and La Frenais 2005: 92–3). Projects supported by The Arts Catalyst onboard parabolic flights can largely be divided into two categories: those that address the human body (such as dance performances by Morag Wightman and Kitsou Dubois, explored further in Chapter 5); and those that address the peculiar behavior of other objects and materials in microgravity: in Vadim Fishkin's *Kapelgraph Zero-G* (2003), drops of water dance around in a transparent container, "changing direction in mid-air according to the subtle variations in the balance forces of the aircraft"; i-Dat's *Constellation Columbia* (2003) was a prototype for a space monument, consisting of six tethered cubes in an arrangement that would reconfigure in mid-air.

The Arts Catalyst initiative arose in part due to the organization's indirect involvement in a 1999 project by Slovenian theatre group, Cosmokinetial Kabinet Noordung, led by Dragan Živadinov, that involved performances in microgravity. *Noordung 1995:2045* is a long-duration performance that began in 1995 on Earth and will conclude in 2045 in Earth orbit. In 1999, the performance was staged in the variable gravity conditions of a parabolic flight, with seven performers and sixteen audience members who experienced the same varying gravity conditions as the performers. This performance will be repeated every ten years at a different site, and as performers die they will each be replaced with a "remote-controlled sign." It is expected that no members will have survived until 2045, so all the performers will have been replaced with objects that can then be launched into geostationary orbit for one final performance (Žiberna 2011: 39). Inspired by the *Noordung 1995:2045* project, Rob La Frenais of The Arts Catalyst initiated the organization's Microgravity Interdisciplinary Research network, which would support further artists to access the microgravity environment onboard future flights (Peljhan 2005: 36).

The influence of The Arts Catalyst has been far-reaching, with other initiatives learning from their approach. Rob La Frenais has advised other groups on their own engagements with microgravity, including *La Gravedad de los Asuntos* (*Matters of Gravity*), for which nine artists and one scientist traveled from Mexico to Russia's Star City in 2014, in order to "explore gravity ... by changing our perspective on it through its absence" (Guzman 2018; Paterson 2015). The works produced or deployed by these artists onboard a parabolic flight included Nahum's *Holding Air* (see Chapter 5, Figure 5.2), Ale de la Puente's *Un infinito* (*An Infinite*, see Figure 3.7), and Tania Candiani's *Machine for Flying, Besnier 1673*, a recorded performance for which Candiani reproduced the failed flying apparatus of the French locksmith Besnier, and re-enacted his flight in microgravity.

In France, CNES's l'Observatoire de l'Espace has established a *Résidence ZERO-G*, which, much like The Arts Catalyst before it, offers artists, philosophers, writers, and researchers opportunities to respond to the weightlessness that they encounter onboard a reduced gravity flight. CNES's call requests that artists avoid proposing tokenistic or "fetishistic" engagements with weightlessness, such as proposals to fly terrestrial artefacts for purely "symbolic" purposes. Instead, it calls for artists to work *with* weightlessness to "evolve the modalities of creation resulting in a "unique object which would be born from gestation in [weightlessness]" (CNES 2022b).[2] The initiative prioritizes the experience of weightlessness, inviting participants to learn from their experiences in ways that will shape their future work. As a result, those selected have tended to focus their attention on the experience of inhabiting a weightless body, including, in 2013, author and director Jean Lambert-wild, who performed choreographed movements throughout the flight, and, in 2022, multimedia artist Renaud Auguste-Dormeuil, who used an array of mobile phone cameras to capture five disoriented views of himself floating through the interior of the aircraft. These videos were displayed simultaneously as *Dansez Maintenant* (2022), as a video installation spread across five screens in an Earth-based exhibition (CNES 2022a; CNES 2023a).

The US-based Space Exploration Initiative (SEI) at MIT Media Lab has a mission to democratize access to space, and seeks to achieve this goal by granting opportunities to scientists and artists, in anticipation of a future in which human habitation extends beyond the confines of Earth. Their vision of a spacefaring future has much in common with the "space

[2]Translated from the original French: "*La visée première de ce projet est l'appréhension de ce qui peut se manifester spécifiquement dans un milieu en micropesanteur. De fait, le projet proposé par l'artiste devra se dégager pleinement de toute approche fétichiste de l'Espace telle que faire voler une pièce existante sur Terre ou jouer une partition écrite sur Terre afin de leur conférer une valeur symbolique après leur passage en micropesanteur. Il ne s'agit donc pas pour l'œuvre d'être la simple transposition d'une élaboration artistique terrestre à une situation d'impesanteur. Tout au contraire, la création attendue est vouée à des contraintes d'existence qui la font échapper d'emblée à la conception classique d'un objet unique qui naîtrait de la seule gestation en atelier. L'enjeu, ici, est bien de construire un projet avec le milieu spatial, et plus particulièrement avec l'impesanteur pour faire évoluer les modalités de création et de réception de l'œuvre.*" (CNES 2022b)

culturalization" espoused by Miha Turšič and KSEVT. SEI's founder, Ariel Ekblaw (2021: xvii) describes an "anthropocosmos" in which human culture extends beyond planet Earth, and the responsibility that humankind has to ensure that spacefaring communities will "reflect the diversity of cultures on Earth." This, she suggests, will not only involve carrying existing cultures into space, but further creating new cultures that may "depart significantly from our earthly assumptions and traditions." Ekblaw (2021: xi–xiv) predicts that collaborations between science, engineering, art, and design in the context of human space exploration will make it possible for new disciplines to emerge. Ekblaw's ambition was to distinguish the SEI from other speculative endeavors by ensuring that the initiative provided real access to the environment of space, primarily focusing on the microgravity environment, which they have accessed achieved by chartering parabolic flights since 2017, and more recently by deploying payloads aboard suborbital flights and the ISS.

The Space Exploration Initiative supported the launch of Ani Liu's *Long Distance Low Gravity Love Letter* (2018) as an untended payload aboard a Blue Origin suborbital flight. Liu's work consists of a payload box lined with paper, housing a piece of graphite (see Figure 3.2). When the payload is weightless, the graphite floats around the box and makes marks on the lining paper, in order to generate a visual record of the journey that it takes when weightless (see Figure 3.3). The gestural marks resemble the work of action painters, and one might consider Liu's work in the context of a history of automatic mark-making and, in particular, mediated gestures of, for example, mechanical mark-makers. The marks might be read as a map or movement-score that records the path of the graphite and its collisions with its enclosure. Liu had also initially intended to capture this data with a sensor that would transmit the data to a robotic arm that could in turn recreate the movements of the payload to a "sister" box located on Earth (Liu and Moran 2021: 60). In deployment, this part of the work did not operate as planned, and there were insufficient data to recreate the payload's movement. The only record of the payloads' journey exists, therefore, in the more poetic form of the graphite "drawing."

As SEI activities have expanded from parabolic to sub-orbital and orbital flights, they have offered researchers a broader range of engagements with differing gravity conditions. Activities onboard parabolic flights provide valuable insights into the potential for creative engagement with the environment of space, but the short duration of weightlessness aboard these flights means that there is still much to learn. Astronaut Michael Lopez-Alegria (2023) stresses that the short-duration weightlessness experienced aboard parabolic flights is not sufficient preparation for the long-term weightlessness experienced aboard the ISS, where lengthy tasks must be completed wholly in microgravity. Parabolic flights offer respite between short periods of weightlessness, during which researchers can take stock and steady or reset their experiments if needed. As sub-orbital flights become more

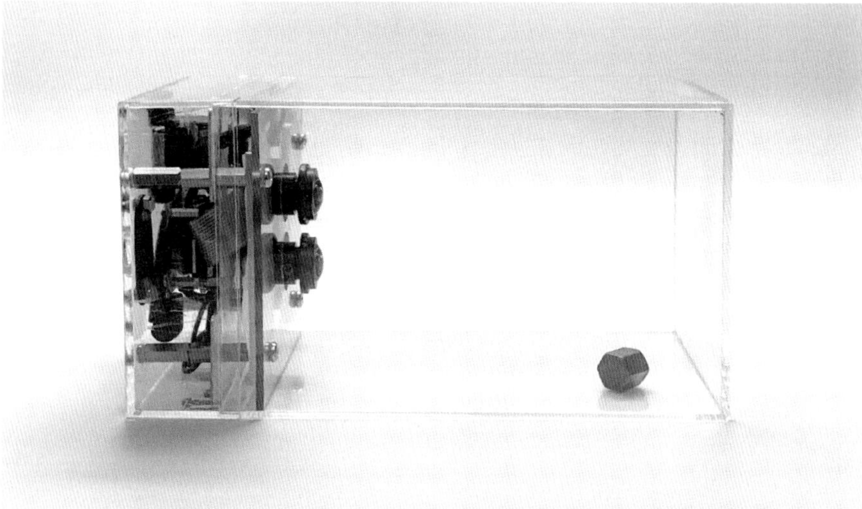

FIGURE 3.2 Ani Liu, *Long Distance Low Gravity Love Letter*, graphite, paper, polycarbonate, microcontroller, camera, aluminum chassis, LEDs, 4" × 4" × 8", 2018. Liu's payload is lined with paper and houses a piece of graphite that acts as a mark-making object as it floats freely around its enclosure in microgravity. Image courtesy Ani Liu.

commonplace, there are opportunities for artists to work in an environment that offers microgravity experiences of a longer duration, but, nonetheless, are still relatively brief (three minutes aboard Virgin Galactic or Blue Origin flights). These short bursts of weightlessness do not allow sufficient time for astronauts to acclimatize to carrying out activities as they will in space. Lopez-Alegria uses the example of typing on a keyboard, and the extent to which this action is reliant on the feet. During the Axiom 1 mission, when pressing the keys on a keyboard he found that the force of his fingertips on the keys would propel his entire body away from the workstation. Despite prior experience of several previous missions to the ISS, he found that he needed to relearn the skill of securing his body by grabbing with his feet, and to maintain that grip throughout the duration of a task. Liat and Segel observed similar difficulties in video recordings of astronaut Eytan Stibbe unpacking *Impossible Object* aboard the ISS (see Case Study 2). The task required Stibbe to maintain constant control of all of the objects in the package, which would drift away if not held in place, while simultaneously attempting to maintain the stability of his own body by grasping onto his surroundings with his legs.

One reason to opt for a parabolic flight or sub-orbital flight, as opposed to a trip to a space station, is the opportunity to experience a variety of different gravitational conditions, and the transition between conditions.

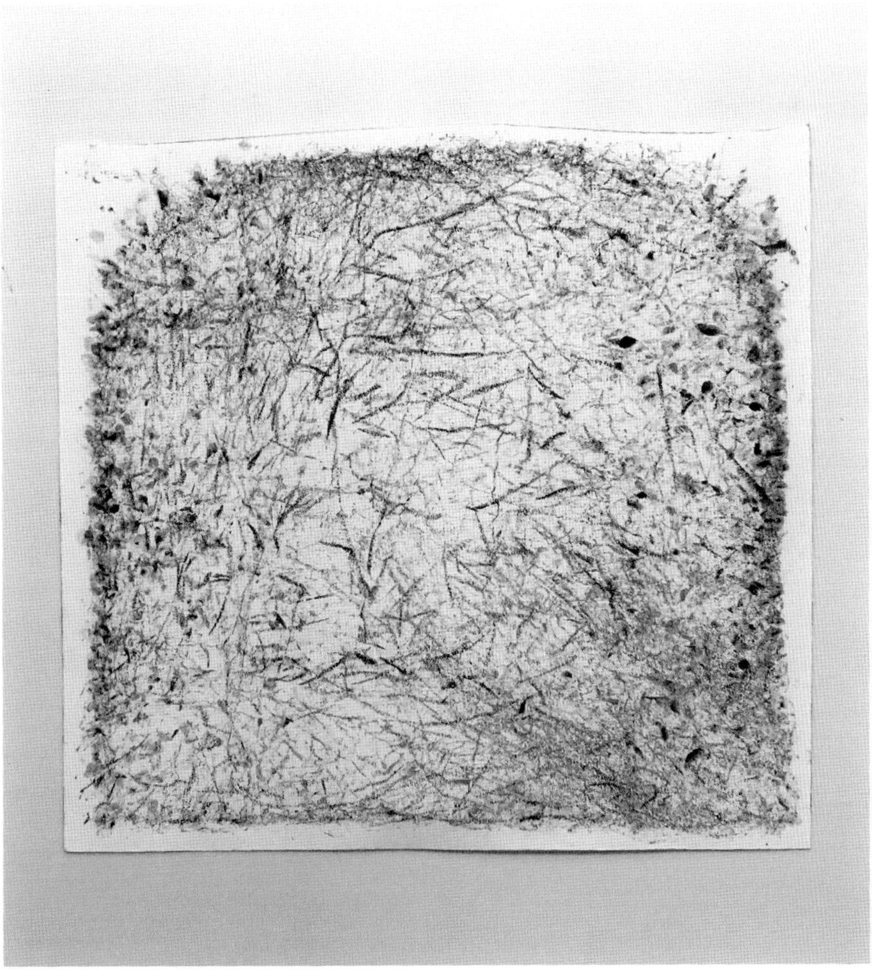

FIGURE 3.3 Ani Liu, telepresent drawing generated by *Long Distance Low Gravity Love Letter* aboard a suborbital flight, 2019. The drawing records the movement of a mark-making piece of graphite that was allowed to float freely within the payload on board a sub-orbital flight. The drawing might be considered as a movement-score that poetically records the movement of the graphite and its collisions with its enclosure. Image courtesy Ani Liu.

Such transitions introduce the possibility of observing an object as it changes from one state to another. Objects and materials can be observed transitioning from their normogravity to their microgravity form or configuration. The transition is quick, but interesting, as objects are reshaped, and appear more fluid than they are when in constant microgravity. For Mikhail Ryklin (2005: 14–15), who experienced both reduced and increased g-forces aboard a parabolic flight, "the transition from the state of weightlessness to the state of double gravity" was "equally important" as the experience of microgravity. For Ryklin, the experience of transition between two unfamiliar gravitational environments reinforced the post-gravity effect, contributing to an enhanced sense that weightlessness and increased gravity are interrelated. In this way, a parabolic flight provides particular insight into the relativity of microgravity.

Various proposals for simulating Earth gravity during long-duration spaceflight have been prototyped onboard the ISS, employing centrifuges that may one day be scaled up to the size of a vessel or space station large enough for human occupation. While this technology is years away from implementation at a scale that may accommodate humans, at a smaller scale it can offer access to multiple gravity environments during spaceflight or in orbit, allowing for direct comparisons between different gravity conditions, and for observation of the transition between them. SEI's *Sojourner* payload, deployed onboard the ISS in 2020, simulates multiple gravity environments simultaneously. The payload models a miniature space station consisting of three layers, each housing six containers for artwork. The payload exposes its contents to three different gravity environments: the top layer is fixed, and so its contents are exposed to the microgravity environment of the ISS; two further layers spin at different speeds to "produce centripetal accelerations that mimic lunar gravity and Martian gravity, respectively" (Liu 2020). Following an open call, nine artworks were selected for deployment on the *Sojourner* payload.

Not all of the works selected for *Sojourner* are physically affected by their gravity conditions, but rather their launch into space is of symbolic significance. Those that do respond to these conditions include works by Kat Kohl, Masahito Ono, and Andrea Ling. Kohl and Ono both took advantage of the microgravity environment of the ISS, concerned primarily with weightlessness, as opposed to the varying gravity conditions offered by *Sojourner* (*Ars Electronica* 2020). Kat Kohl's (2020) *Memory Chain: A Pas de Deux of Artifact*, a sculpture consisting of two chains, one extracted from the artists' grandmother's watch, and the other from their grandfather's tie clip. When weightless, the two chains "dance retelling the story of this romantic partnership. Masahito Ono's (2020) *Nothing, Something, Everything* includes a small, cylindrical magnet that, when weightless, points toward the Earth's magnetic north. Ono's work simultaneously responds to the distance from, and proximity to the Earth, exploiting weightlessness to allow free movement of the magnet in its container, and the pull of the Earth to orient that free-floating object in space. The work is a reminder that spaceflight takes humankind beyond the confines of Earth, and yet is still engaged in Earth-centric activity, such as orbital flight. Where the *Sojourner* works are spread across the three levels of the payload, and hence exposed to different gravity conditions, the artists' intention is not to observe the effects in real-time but to exploit the long-term changes that these environments prompt within the work, informing later activity on Earth. Andrea Ling's

(2020) *Abiogenetic Triptych* is formed of three parts across the three layers of the payload so that each part is exposed to a different gravity condition. The three containers include mixtures of inorganic salts, representing the conditions in which new life forms may emerge. Once these containers were returned to Earth, the salt mixture had each formed different structures in response to their different gravity conditions, and were used by the artist to inform speculations about the kinds of organisms that might emerge under different gravity conditions.

Representing microgravity on Earth

The location of audiences is significant for any postgravity art as it dictates whether the sensation of weightlessness becomes part of their experience of the work. Sixteen audience members who flew with Cosmokinetial Kabinet Noordung aboard its parabolic flight were able to observe the microgravity performance of *Noordung 1995:2045* while simultaneously experiencing the weightlessness of their own bodies. Michael Benson (1999), who was onboard to record the performance, describes the experience of weightlessness "as if a magic wand is waved, and the most fundamental physical rules are suspended."[3] So novel was this experience that "the sensation of weightlessness itself took precedence ... over the actions of the actors." In this way, director Dragan Zivadinov was able to "break down the experiential barrier between audience and spectator" to such an extent that, for Benson, it led to total identification with the actors. Benson records feeling that he "became an incontrovertible part of the piece of art [him]self."

This opportunity is not typically available to audiences of art performed or deployed in microgravity. In what Annick Bureaud (2003) describes as "an ultimate state of elitism," the only "direct" audience is often a small handful of astronauts and cosmonauts. Work deployed in microgravity is often disseminated to larger audiences on Earth through recordings, descriptions, and duplicate displays. CNES's call for microgravity projects notes that audiences for these works will be on the ground, emphasizing the return to Earth and subsequent exhibition in order to encourage, enable, and promote cultural and creative uses of space to audiences that are larger and more diverse than those who have the privilege of experiencing microgravity first-hand. Other initiatives and artists have done the same, recognizing that audiences will not likely be located in the weightless environment when

[3]Since this performance took place onboard a parabolic flight where several different gravity conditions are experienced, Benson was also able to experience first-hand the transition between conditions: "What had been three-dimensional, and kinetic, and weightless is now being pushed down like sculptors' clay on the floor, immobilized, denting that foam-rubber in an overpowering roar of turbo-jet engines."

viewing the work, but will observe the effects of weightlessness through records made during the flight. The nine artists whose works formed *La Gravedad de los Asuntos* (*Matters of Gravity*) had their works subsequently exhibited in a traveling exhibition that began in Mexico City in 2015. Similarly, artists who produced work for the *Moon Gallery* were required to produce duplicates of the work that was transported to the ISS, so that those duplicates could be exhibited on Earth and sold through auction to support the project.

It is perhaps as a result of the knowledge that their work will later be exhibited on Earth that many works in microgravity are principally concerned with recording those visible, recordable effects of weightlessness. Ale de la Puente's *Un infinito* (*An Infinite*), for *La Gravedad de los Asuntos*, records the effects of weightlessness on an hourglass, "reflecting on the links between time, space, and gravity." The video shows the sand of the hourglass pooled in its bulbs, with each bulb containing its allotted volume of sand in perpetuity (see Figure 3.4). The sand does not appear to be frozen in time, as if the flow of sand has been paused, but rather it is as if this flow never has and never will happen: the possibility of flow seems to have been extracted from the hourglass, as if it is *without* time. De la Puente's video, along with others from the mission, was exhibited in a traveling exhibition that began in Mexico City in 2015. For the audiences of this exhibition, de la Puente's work is a visual record of the effects of weightlessness that they will likely never experience with any sense other than sight. In other examples, efforts have been made to record or represent weightlessness both visually and sonically. For The Arts Catalyst project, *Wave/Particle* (2001), Ansuma Biswas and Jem Finer sealed a camera and microphones inside a crate in order to record the behavior of liquids sealed inside transparent cubes, alongside sounds made by a Baoding ball as it struck the walls of the crate. Following the flight, the video recording of these objects was presented as triptych to audiences on Earth.

There is still work to be done identifying who the Earth-bound audiences for spaceworks may be, and how microgravity works should be best disseminated to them. Max Baraitser Smith (see Case Study 3), whose work was selected for the *Moon Gallery*, notes that there was a common understanding shared by members of the team that there would be a "lack of predefined audience" for the project. There would be no one other than a few astronauts to view the work onboard the ISS, and similarly few lunar astronauts to witness it at its final destination. With the field of space art in its youth, there is no established audience that would seek out video of the work, nor an established framework for disseminating or streaming live or recorded footage, and Baraitser Smith and his peers felt that the work might lack purpose if viewed by those astronauts alone. Until such time as larger audiences are familiar with space art, they will not seek it out, and so showcases are necessarily exhibited under the umbrella of broader or tangentially related themes, where audiences may have a greater or lesser degree of sympathy for the goals of space artists.

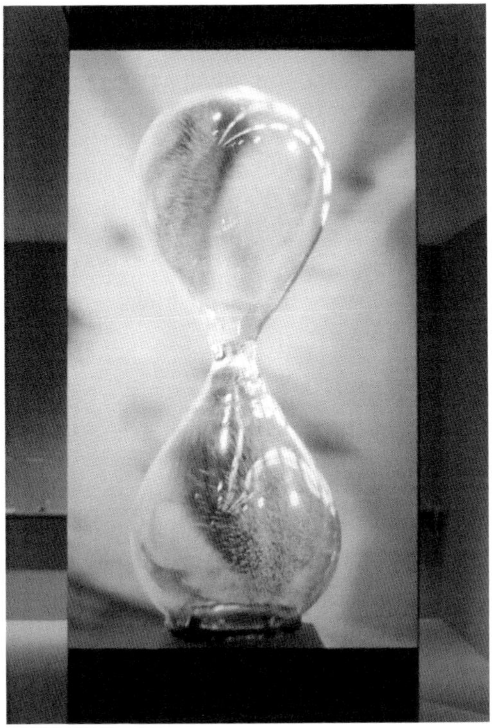

FIGURE 3.4 Ale de la Puente, *Un infinito*, silent vertical video (still), 9:16, 2:22, 2015, displayed in a terrestrial gallery setting. Recording the effects of microgravity on an hourglass, de la Puente's video shows sand pooling in each chamber. While the flow of sand would, in normogravity, measure the passage of time, in microgravity the sand appears to defy time; swirling in perpetuity. The video was exhibited to audiences on a vertically oriented screen as part of a terrestrial, traveling exhibition, in normogravity conditions, giving audiences a visual experience of the effects of weightlessness. © Ale de la Puente

The choice of location for Earth-based exhibitions or showcases dictates the nature of the audiences and the messages that such activities will communicate. Space exploration organizations have a well-established history of engaging with audiences through Earth-based public exhibitions of models and actual artefacts that have returned from space, typically in museums. Science museums such as the London Science Museum, space museums such as the Moscow Cosmonaut Museum (Музей космонавтики), and programs such as the NASA Johnson Space Center's Johnson Exhibits Program grant access to scale and full-size models, replicas, duplicates, and objects that have returned from space in order to educate and engage the general public. The educational and cultural value of viewing artefacts that have been visibly altered by the environment of space is one of the most important ways in which lay audiences are helped to understand space and spaceflight, and the environments encountered in space. The Soyuz capsule on display in London's Science Museum bears the scars of re-entry, and so can be presented not just as an engineered object, but also as a storytelling device, its condition manifesting the risks that ESA astronaut Tim Peake took during his return from the ISS in 2016. Worn and damaged artefacts such as these have an authenticity that cannot be matched with replicas, and their physical condition enables narratives that position them not only as inanimate objects but also as witness to historical events.

By displaying spaceworks returned from space in these same contexts, artists run the risk of their works being "fetishized," a risk that CNES cautions against (CNES 2022b). Positioning spaceworks alongside artefacts of space science or space exploration reframes them as "symbolic" and thereby diminishes their value to the arts, as works of art rather than evidence of spaceflight. Nonetheless, CNES does request that artists taking part in its *Résidence ZERO-G* program do make their work available for exhibition following their flight, suggesting that Earth-based showcases might help more diverse audiences to recognize the potential for art in space, and to begin to understand the effects of weightlessness on arts practice. For these exhibitions, CNES favors an art gallery setting where the works can be approached primarily as artworks as opposed to artefacts of spaceflight. In order to avoid fetishization, and to further the goal of conveying to audiences the value of enabling artists to directly experience spaceflight, it is vital to not only display the spacework but also to represent, as authentically as possible, the environment in which the work is intended to be encountered.

When work is displayed or represented to an Earth-bound audience in a gallery or museum setting, there are methods of enhancing the audience's experience in an attempt to replicate, or at least better represent, the experience of weightlessness within the gallery space. These methods are used to ensure that the exhibition represents not only the spacework as an artefact, but also the embodied experience of weightlessness that audiences are denied by being Earth-bound. This may involve, for example, rigging the work so that it floats as if weightless in the exhibition space. The London Design Museum's *Moving to Mars* exhibition (October 18, 2019–February 23, 2020) featured artefacts strung on wires, as if floating. The exhibition suspended Anna Talvi's *Antagonist Exo: Muscle Bodysuit* (2019) in a rigged display, featuring mannequins suspended from the ceiling at a variety of angles, so that visitors to the exhibition encounter Talvi's design from below. The display mirrors the angles at which one might encounter a weightless body during space travel. Similarly, Xu Bing's terrestrial exhibition of *卫星上的湖泊* (*Satellite Lake: Cosmic Reflections* 2021–22), features video footage of the work filmed in orbit against a backdrop of the view of the distant Earth. Audiences in the Red Brick Art Museum, Beijing, were invited to view this video from below, looking upwards at the projected image as if both the audience and screen were displaced by microgravity (Bing 2022).

In and beyond the gallery space, spaceworks can be shared with terrestrial audiences in photographs and video recordings that document the work's deployment in space. When weightless artefacts are viewed remotely or in a recording, weightlessness is reduced to a visual experience rather than a bodily experience. It is, however, possible to enhance visually mediated experiences, making them more immersive and sensorial, using carefully considered approaches to the staging and filming of an object. Adriano D'Aloia (2012) suggests that at "the very core processes of film 'viewing' [is] an experience of tangibility or more accurately in representations of weightlessness, the "intangibility" of being suspended "in the void." D'Aloia explores how the sight of a subject floating in space invites audiences to imagine the bodily experience of weightlessness, as if experienced vicariously through the subject. Beyond simply representing the weightless environment, camera motion can make filmed footage more sensorial, more immersive, as if transporting the viewer into the weightless environment. Vivien Sobchack (1982: 320) suggests that "the camera is originally understood as inhabiting and expressing space humanly rather than mechanically. It is perceived existentially and experientially as *bodily perception* and *expression*." The weightlessly floating camera therefore signifies the weightlessly floating camera operator, and, more broadly, the weightlessness that can be experienced by the human body. In this way, the viewer is invited not simply to observe the weightless object on screen, but to imagine occupying the same space as that object.

Recalling his experience of filming *Noordung 1995:2045* onboard a parabolic flight, Michael Benson (1999) describes how, "the camera ... suddenly weighs nothing at all, and leaps from the hand like a bird" then drifts away, continuing to film, "recording the bizarre scene all by itself, like a slowly-spinning autonomous satellite." The weightless camera, then, "can move through its own trajectory," and, suggests Benson, "you could almost say that the camera itself also sought to become an actor. An autonomous astronaut" (*sic*). The cameraman's loss of control, and the camera's autonomy, can be artificially simulated on Earth in order to create an artificial weightlessness effect for authentic representations of the microgravity environment. A microgravity effect is produced in science fiction in the television series, *The Expanse*, in which the quality of camera motion is used to create the impression of different gravity conditions. The show's cinematographer, Jeremy Benning, describes how weightlessness can be conveyed "by having the camera slowly rotating on its axis" (Sim 2020). Benning reviewed many hours of NASA ISS footage and observed that, when these cameras are released from their rigs, there is a slow, subtle drift and rotation that reveals the camera's weightlessness. As weightless objects rotate around their own center of gravity, this rotation appears to originate from the nodal point (the center of the lens). Benning therefore sought out a variety of cameras and riggings that would allow motion and rotation that originates from the nodal point. Benning describes this kind of camera motion as "nodal float." There is much that can be learnt from Benning's approach where artists aim to represent, on screen, not just spaceworks, but also the weightless environment, so that audiences may better understand the significance and intended environment of a spacework. Where footage is filmed onboard the ISS or in any other weightless environment, it may be beneficial to forego the option of fixing the camera's position in that environment, in order to take advantage of the weightless camera's drift and rotation, the more effectively to represent the environment in which the work has been deployed.

All these methods, of course, are necessary only at this particular moment in the history of space exploration, in which the audience is distanced from the artwork as a result of the exclusivity of spaceflight. As space is democratized and access to microgravity becomes more widespread, the question of how to exhibit spacework on Earth will be replaced by the question of how to design galleries in space. This question is, for now, so speculative that it is beyond the scope of this book, but will involve considerations such as orientation and rigging that are explored in the next few chapters.

Case Study 4
Lisa Pettibone, *Verdant*, 2022

Verdant is a small, flexible sculpture of translucent glass beads, strung together in a four-legged configuration that is designed to reconfigure in the weightless environment (Figure 3.5). The work was deployed to the International Space Station in 2022, where it was exhibited alongside 63 works by other artists in the *Moon Gallery* (see Chapter 2). The 2022 deployment onboard the ISS is a test-flight in preparation for the eventual deployment of a similar gallery on the surface of the Moon. Although future iterations of the *Moon Gallery* will be in Moon gravity conditions, Pettibone chose, for the work's first deployment, to engage with the microgravity environment that exists onboard the ISS.

Choosing weightlessness as her primary concern, Pettibone approached the design of *Verdant* with an awareness of the lack of a stable paradigm of orientation that exists in microgravity, and the subsequent potential for her work to lack a definitive configuration or resting point. As a result, the piece has no designated top or bottom, and presents as a different shape

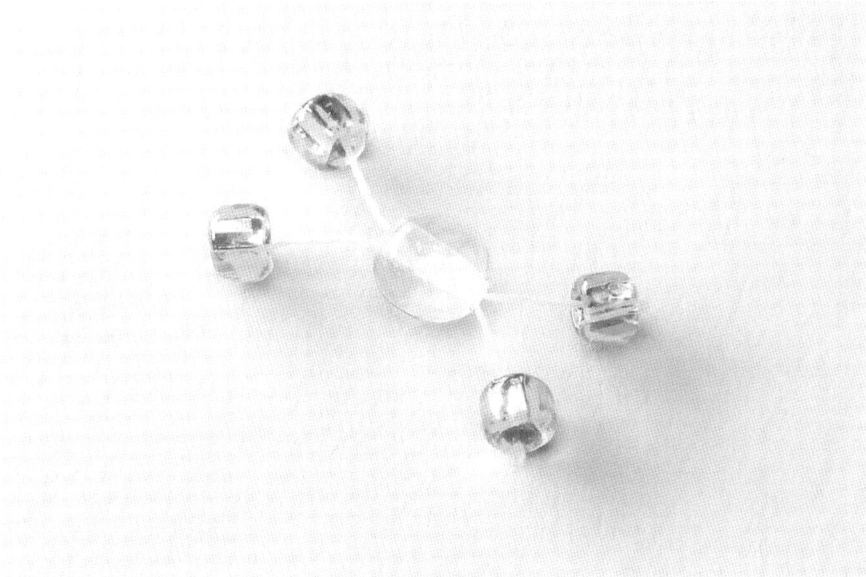

FIGURE 3.5 Lisa Pettibone, *Verdant*, glass and thread microgravity sculpture, 15 × 8 mm, 2022. Pettibone's miniature sculpture is based on the central core of the chlorophyl molecule (1 clear magnesium atom surrounded by 4 green nitrogen atoms). The sculpture responds to weightlessness by altering its form. The tethers, much like umbilical cables that tether astronauts to their vessel during a spacewalk, anchor the glass beads to one another, creating an envelope of potential movement. © Lisa Pettibone.

when viewed from different angles. Contributing to the work's potential to transform is the flexibility of the threads that anchor the beads to one another. The configuration of the piece may be different from one moment to the next as it floats weightlessly in the microgravity environment of the ISS and the beads move in different directions. In Earth gravity (as depicted in Figure 3.4) the beads adopt a flat configuration, as gravity causes them to rest on a planar surface. In microgravity conditions, these beads may move independently, so that the overall form is more fluid and fluctuates in three dimensions.

On Earth, gravity and friction constrain the movement potential of an object. However, in space, objects move more freely, and therefore the space sector finds it necessary to artificially restrain objects or their parts. Methods of restraint employed on space stations have ranged from Velcro, to clasps, cleats, and tethers. The process of designing the object involves finding a desired balance between freedom and constraint, arising from a decision about the extent to which the object should be allowed to respond to weightlessness. In *Verdant*, beads are flexibly tethered to one another, able to move freely within a range of movement dictated by the length of the threads. Parallels can be drawn with the range of motion and extent of an astronauts' reach when performing tasks in space, known by NASA as the "work envelope" (Benson et al. 2019). By tethering the beads to one another in the production of *Verdant*, Pettibone has dictated the extent of their potential range of movement. This process did not involve designing a single, fixed configuration, but rather enabling a range of potential configurations.

Verdant is the next logical step in Pettibone's two-decade-long exploration into the ways that materials are shaped by gravity. Her journey began when she became "aware of the transforming effect gravity could have on [her] work while peering into a small test kiln." She describes watching as "a narrow slice ... in a small glass cube began to fall away from the body of the block. It was moving not just because of the extreme heat ... but also due to the natural forces pulling it down. The cooled piece became a record of a battle between material structure and gravity." These observations led to *Weightless* (2005) a sheet of glass shaped by being slumped over an arrangement of cones, displayed so that it appeared to be "floating along the woodland floor." By allowing glass sheets to fall over cones their resting form could be shaped by gravity as it pulls the hot glass toward the Earth. Once rigid, the forms could be turned over, so that they appear to defy gravity and reach upward, toward the sky. When turned upside down, Pettibone observes, the "perception of a piece changes radically." The natural folds appear to defy the forces that made them, billowing outward, against the pull of gravity.

In this and subsequent works, Pettibone has attempted to capture the "energy" of falling (see Figure 3.6). In her process of "free-fall slumping she allows hot glass to spill over and beyond her molds, so that it appears to have become captured in the process of falling. She finds that "when an object [adopts] the mold shape completely, it becomes dead," whereas, when

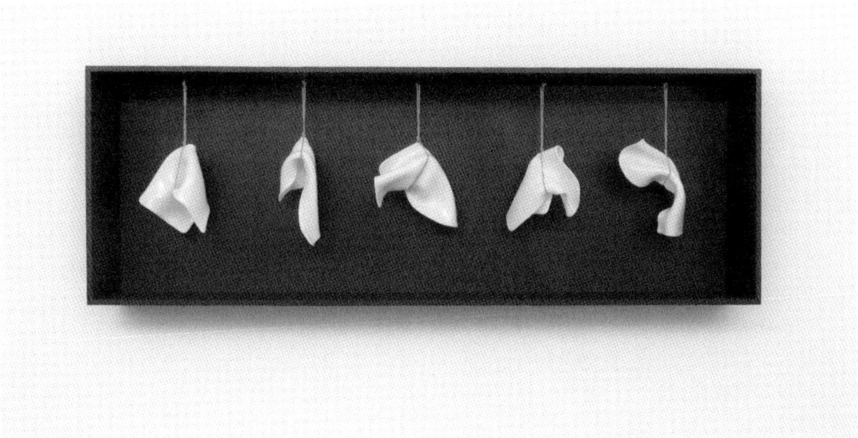

FIGURE 3.6 Lisa Pettibone, *Five Days of Falling*, Slumped glass, fluorescent elastic cord, wooden box, 50 × 148 × 15 cm, 2020. Pettibone's Earth-based work explores how materials are shaped by gravity. The artist describes *Five Days of Falling* as an "ode to astronauts in space." The red cords tether the slumped glass to their container much like astronauts undertaking spacewalks are tethered to their spacecraft by umbilical cables. © Lisa Pettibone.

the glass misses the mold, it appears to preserve the energy of falling. Where possible, she tries to maintain the curves that are acquired through falling, leaving them "uninterrupted" in the final form. There is, in that final form, "a kind of flow" that exists only in objects shaped by gravity.

Having initially acquired experiential knowledge of the effects of gravity through direct observation of its effects, Pettibone later began to explore the science of gravitation, finding that this new knowledge caused her to see the world, and her practice, in new ways. The scientific understanding of gravity led Pettibone to an awareness of the different ways that gravity shapes objects on Earth and in space. While on Earth, and in Pettibone's previous work, forms are pulled toward the Earth, shaped by the pull of the ground; in space, there are "forces coming in all directions." She came to appreciate that gravity is not always a monodirectional force.

Pettibone's understanding of gravitation has evolved through reading and conversations with scientists, but she stresses that, as an artist, "the experiential knowledge is first." First and foremost, she "understand[s] material as a sensory experience." Science, like any new knowledge, enriches Pettibone's practice, but she found that she "didn't need to understand the science deeply." She has found conversations with scientists more useful than scientific literature, as their verbal descriptions have tended to be more "descriptive, and expressive." "Just a little bit of understanding can go a long way," she says, changing the way we see our surroundings in profound ways. She uses the example of learning about how the Moon is falling toward

Earth but accelerating away from us at the same time. Although she has not acquired a mechanical or mathematical understanding of this phenomenon, the broad awareness has inspired her to consider the Moon in new ways that will inform her art. "Anything that changes your perspective," she says, "can give you ideas."

The two types of knowledge that have informed Pettibone's work illustrate challenges encountered by any artist hoping to produce work for space in these early days of the commercial space age. The experiential knowledge that has informed Pettibone's engagements with gravity has been acquired through direct observation and experimentation. As explored in Chapter 2, "experiential knowledge or tacit knowledge as an integral part of [art] practice" (Niedderer and Reilly 2010). Much of the knowledge that informs creative practice is not formally recorded or recognized, but implicitly understood. Familiarity with the effects of gravity, gained through a lifetime of operation within Earth's gravity conditions, is one of the fundamental, if unspoken, principles on which art practice is founded. Opportunities to gain experience of the effects of weightlessness are scarce, thus artists creating work for space are more reliant on abstract knowledge of the effects of weightlessness, as communicated or recorded by third parties. This introduces an element of uncertainty to the deployment of work in space. There are no opportunities to experiment with the behavior of materials prior to deployment, as the artist typically has only one opportunity to

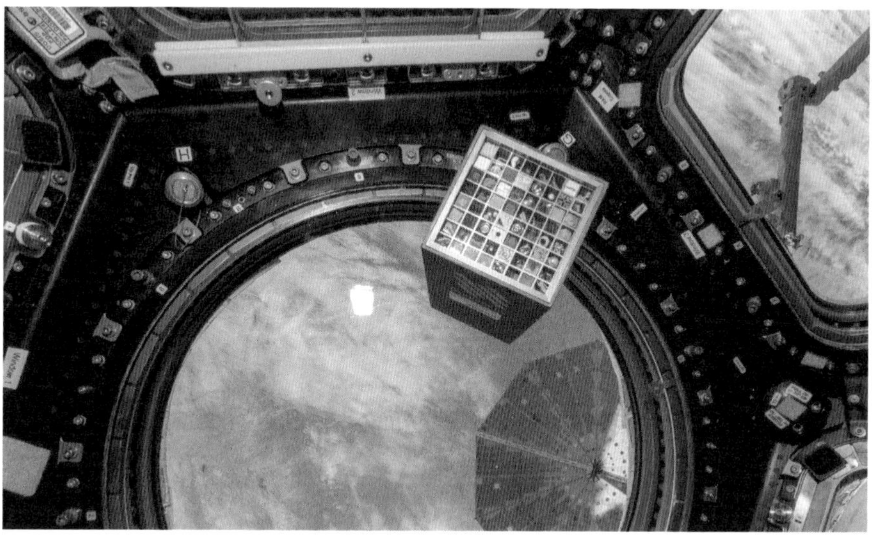

FIGURE 3.7 *Moon Gallery*/Nanoracks unit onboard the ISS in March 2022. On one face of the cube there are sixty-four 1 cm^3 compartments, each containing a work by a different artist, including Lisa Pettibone's *Verdant*. The cube's other faces display a list of the contained works and contextualizing text. Image © NASA

launch a work into space. "You only have one shot says Pettibone, and "you've got to get it right."

Having no opportunity to experiment or test prototypes has meant that, by necessity, *Verdant* is "a very experimental project." In order to predict how the work might behave in weightlessness, Pettibone conducted practical experiments in her own studio, picking up the sculpture with tweezers and turning it around to test the range of possible configurations. Visual records of the mission suggest that the work did not move as much as Pettibone had intended, or as much as other objects in the gallery. The reconfiguration potential of *Verdant* is dictated by the length of the thread arms that tether the beads to one another, as well as the 1 cm^3 size of the cube in which the work is contained (see Figure 3.7). For Pettibone, the size of this container was significant not only because it restricted the size of the work, but also because it limited the potential for the work to respond to weightlessness. With so little space surrounding the work, it was not possible for Verdant to achieve the full range of potential configurations that it might have adopted if floating more freely in a larger space. In hindsight, Pettibone suggests that she might have "made the arms a bit shorter," but acknowledges that they would have not articulated in the same way if this had been the case.

The ways that *Verdant* was restricted by the small size of the *Moon Gallery* compartments illustrates the extent to which, at this stage in the development of the commercial spaceflight industry, it can be necessary for artists to compromise to see their work deployed in space. While the industry has matured sufficiently to offer Pettibone and others the opportunity to see their work deployed, it has been necessary to operate within severe constraints. The lack of opportunity to travel with the work into space further limited *Moon Gallery* artists' opportunities to observe and reflect upon their work in its intended environment. Pettibone expresses disappointment that there is no record of ISS astronaut's responses to her work. The video and photographic records of the work onboard the ISS provide limited evidence of the ways in which *Verdant* behaves while weightless, or of the way that the light conditions of space affect the appearance of the green-colored beads. Astronauts' commentary, along with more detailed recordings of the work in-situ, would have been valuable in expanding Pettibone's knowledge of the effects of the environment of space on her work, and hence better enabled her to adapt and enhance her work for deployment on any future missions.

Lisa Pettibone was interviewed on January 27, 2023.

4

New Horizons

The view from above

In space, there is no up or down, top or bottom. Space is not located above the Earth; rather, it is around the Earth, and space travel invites us to discard the myth that *up* and *down* are universal concepts. Space stations do not have floors or ceilings, and objects do not naturally rest on any face or axis. Astronauts experience bodily disorientation and rely on visual cues to direct themselves into alignment with their workstations. In order to create objects for this environment, it is necessary to revisit what we have learned about directionality, and to imagine our bodies and the objects we create as if they are globes revolving around their own axis.

Living on Earth, humankind has developed a paradigm of orientation, or "culturally standard axes," within which "the flat surface of the world" is the standard for the horizontal axis (Dilworth 2005: 151). Our common paradigm of orientation tells us that the phenomenon we understand as "upright" is orientation in the direction of the sky, and "upside down" is orientation in the direction of the Earth. Vertical orientation, thus, "coincides with the pull of gravity" (Arnheim 1960: 69). As a result of inhabiting this environment alongside objects that are shaped by the same gravitation, we perceive objects as having top and bottom ends or faces, and assess objects' orientation relative to the agreed paradigm of orientation. Human exploration of space has challenged this paradigm of orientation, exposing humans to views of the Earth from above, and of the globe in its entirety. Aerial and satellite images have given humans a "God's-eye view" (Steyerl 2011). Space travelers see the globe in the round, and experience life without orientation.

Progress toward a non-oriented concept of space began with the loss of the horizon. The horizon acts as a visual cue to orientation, demarking the *horizon-tal*. As such, it plays an essential role in our understanding of Earth-bound existence within a stable paradigm of orientation, where land is beneath us and sky above us, where life is bound to a flat surface and grows perpendicular to this plane. Landscape art employs the horizon to separate the land, that is inhabited by the artist, from the sky above. In this way the horizon divides the human world from the heavens: it is "the junction between terrestrial and celestial domains" (Freedman 1997: 181). The horizon is "visible yet illusory, quite real yet nonexistent" (Kochetkova 2020: 576). It "defies being taken as an object due to its evasive character." As we move toward the horizon, it withdraws (Dahl 2010: 237–8). Space is located beyond the horizon, and remains perpetually out of reach. Dancer Jeanne Morel (see Case Study 7) describes a continuing history of art in which sky, that has typically been depicted as a backdrop, can now be foregrounded. Throughout the history of art, "sky is a subject," suggests Morel, and artists now have an opportunity to "continue this history of art" by directly exploring and inhabiting sky "as an environment."

The Western tradition of linear perspective arises from, and has contributed to, a conception of the Earth as a flat plane. In linear perspective, "the curvature of the Earth is typically disregarded. The horizon is conceived as an abstract flat line upon which the points on any horizontal plane converge" (Steyerl 2011). Air and space travel have granted access to views of the curvature of the Earth. When a spaceflight participant travels away from the ground, they witness the extension of the horizon from a line to a curve. The changing horizon is part of the overview experience, evidence of the curvature of the Earth. The further the spaceflight participant travels from the Earth, the more the horizon curves, so that the curvature reflects their distance from Earth's surface. Lunar astronauts have traveled so far from the Earth that they have seen the horizon curve in on itself and meet at the other side of the globe, becoming a complete circle. The Earth from this distance is not a surface, but a container, moreover, a finite container in which all known living things reside.

On the Moon, a new horizon is formed by the lunar surface (see Figure 4.1). The presence of this alien horizon has the effect of othering the Earth. While the lunar surface in this image is close and knowable, the Earth is distant and strange. The Earth's familiar surface features and atmospheric phenomena are abstracted to shapes and swirls. This image reinforces the idea that the space beyond the horizon is always unknowable and unreachable, and that the ground below the horizon is a location of residence, albeit temporary residence. Alien landscapes, when located nearer than the horizon, lose some of what makes them seem extra-terrestrial. While they may retain some of their strangeness, the apparent location of the observer on the ground before the horizon is evidence that they have been discovered and inhabited by human bodies. Thus, they become part of the expanded territory of Earth.

In aerial view, the horizon is absent altogether. If the horizon is understood to represent a horizontal axis, then, where the horizon is absent, there is room to imagine an artist or observer who is differently oriented. Horizontality is exposed as an Earth-bound concept, and the potential for space to be reconfigured becomes apparent. Familiar landmarks and features are abstracted. The Earth becomes a series of lines, shapes, and patterns. The aerial view grants new insights into the formation of the landscape,

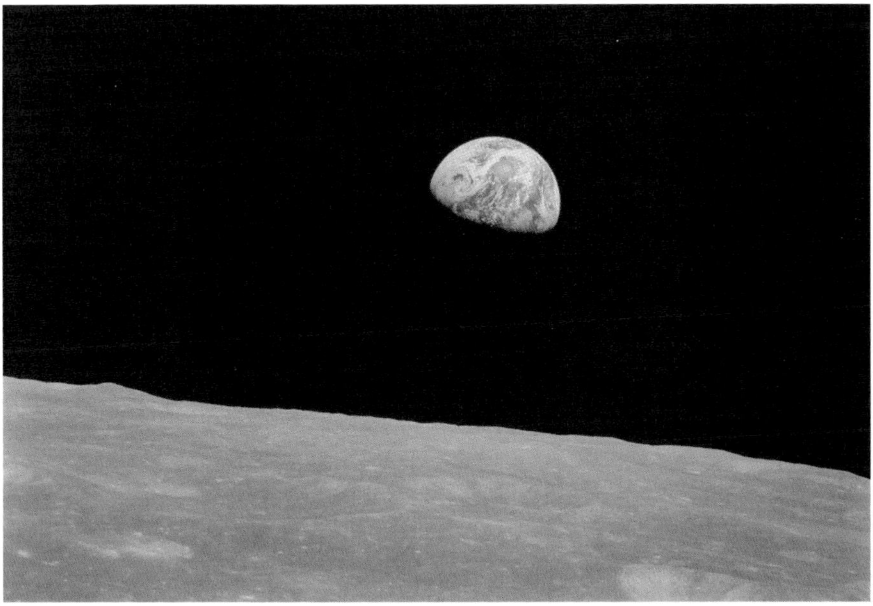

FIGURE 4.1 Apollo 8 Mission image, Earth rising over the horizon of the Moon, 1968. A horizon establishes the location of the photographers—the crew of the Apollo 8 mission—in close proximity to lunar surface. The Earth appears in the sky as a celestial object. © NASA

including interrelations between features of the landscape that are not visible in ground-level viewing. Even in aerial view, the sense of ground is reinforced. Despite these images positioning the audience in the space above, they direct attention toward the Earth. The sky is absent from the image, and yet, at the same time, these images are only possible by inhabiting the sky. Thus, these images of the Earth seem anything but Earthly.

The progression from landscape to aerial view represents a shift from the land as it is experienced and the land as it is represented. Astrid E. Schwartz (2009: 37–40) suggests that the aerial view of a landscape has scientific connotations. The aerial view is accessible only as a result of the introduction of aerospace technologies, and, prior to this, in cartography, site views, and panel paintings. The aerial or cartographic view is how, suggests Schwartz, landscape entered the sciences, having been, in the arts, "subjective … aesthetic," becoming, in aerial view, "objective, transparent and clear-cut." Stephen Graham (2016) suggests that representations of the Earth as a flat plane have made it too easy to establish and divide territories, for borders to be abstracted, ignoring the "three-dimensional politics of the worlds above, below and around borders." Drones and satellites, at an altitude so high that they can traverse those borders, capture images that are "often invoked by states as evidence of unparalleled veracity," used to justify border incursion at ground level (Graham 2019: 210).

Aerial views reveal the potential for the Earth to act as a canvas. Artist Pierre Comte (1987: 17) imagines the potential for "Earth/Space dialogues" in which manmade formations on the Earth's surface can be read as signs by distant space travelers. *Pacific Triangle* is one of several works proposed by Comte that would, if realized, form a dialogue between the Earth and Space. His large-scale works of "Planet Art" are intended to be realized at scales greater than any Earthworks made previously, and function as "terrestrial installations directed to the cosmos" (Comte 1987: 18). By "cover[ing] the ground with 'signs' for the astronauts," Comte would mark the planet as inhabited by humans, set apart from its neighbors in the solar system not just by its ability to sustain life, but also more specifically as a planet that has been shaped by a species—the human species—that seeks to communicate with the wider cosmos. These ideas have been manifested in work such as Eduardo Kac's *Lagoogleglyphs* (2009–present), large-scale glyphs displayed facing toward the sky so that they may be captured in satellite images and viewed via Google Earth (see Case Study 7). Annick Bureaud (2004) suggests that this kind of work, that exists within the context of the "shift in relationship of scale between us and our planet" that we might associate with the Overview Effect, goes "beyond Land Art" and uses "Earth as a medium."

Observing the Earth from above and from afar, we are reminded of the extent to which humankind has become a "post-Earth" species. Views of

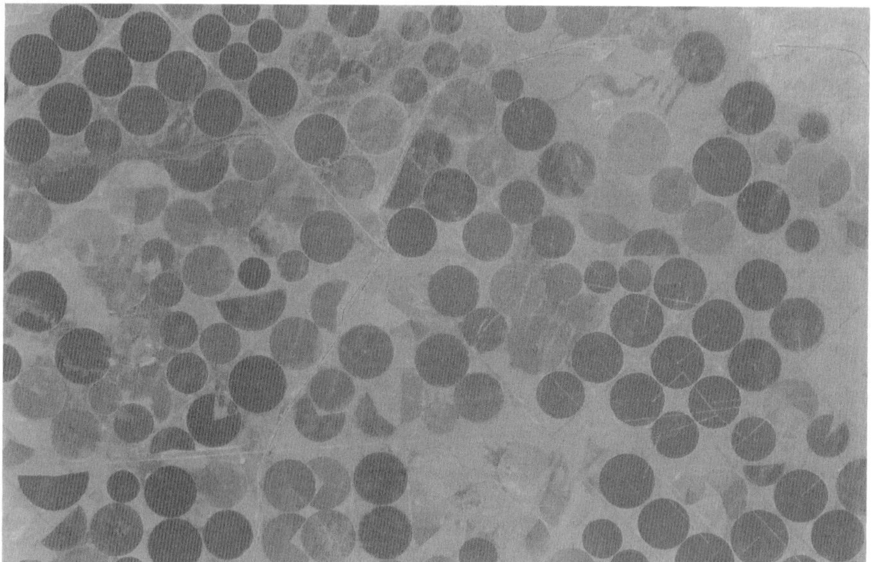

FIGURE 4.2 Earth Observation taken during a day pass by the Expedition 40 crew aboard the ISS, 2014. Observed from above and at a distance, humanmade features of the landscape are abstracted, in this instance becoming more pattern than object. © NASA

Earth from a distance lead to an experience of "world alienation" in which our home planet no longer seems familiar (Du Preez 2019: 83–4). In a post-Earth existence, we are liberated from many of the systems and conditions that shape our everyday experience on Earth, including the experience of being bound to Earth's surface, with ground below and sky above. We overcome established ideas about "down" and "up," recognizing that the Earth can be both and neither above and below us.

"There is no upright in space": Abandoning what we know about orientation

As we move beyond Earth, we abandon the fixed paradigm of orientation that has arisen as a result of our Earth-bound existence. Spaceflight inevitably challenges the primacy of the upright, and "the conventions that surround the human experience of the vertical dimension must inevitably break down" (Graham 2019: 206). In order to engage with the environment of space, it is necessary to appreciate the extent to which notions of verticality are peculiar to planet-surface habitation, and that no space, nor object contained within it, must be naturally oriented in alignment with the pull of Earth gravity. Images captured onboard the ISS illustrate the ways in which orientation is relative and subjective in space, the extent to which astronauts operate without common orientation, and the extent to which orientation is imposed on their environment. In Figure 4.3 astronauts pose for a photograph at all orientations relative to the camera. The frame of the image suggests an intrinsic orientation, but those astronauts who are oriented counter to this do not fall to the floor, nor do they exhibit any visible signs that we might associate with being upside-down on Earth. They are just as comfortable as the astronauts who are posed at the same orientation as the camera.

Cosmonaut Valentin Lebedev describes the need to readjust to life without common orientation, finding that objects and the procedures for interacting with them are defamiliarized when the objects are encountered at unfamiliar angles. When boarding the Soviet space station, Salyut 7, Lebedev recalls, "I was surprised that I didn't recognize it, maybe because I came in at an unusual angle—I walked in on the side panels." The actions required to enter the space station were the same as he had rehearsed in training, and yet "we tried to open the hatch of the space vehicle, but we couldn't. I was only able to do this by standing upside-down" (*sic*). Once onboard, Lebedev found that everyday tasks were made more complex because the intended orientation of objects could not be taken for granted:

> You turn up-side down, everything looks different. For example, you take the frame of the equipment and try to put it in what seems to be the right place, but it won't fit. So you start rotating things in your head,

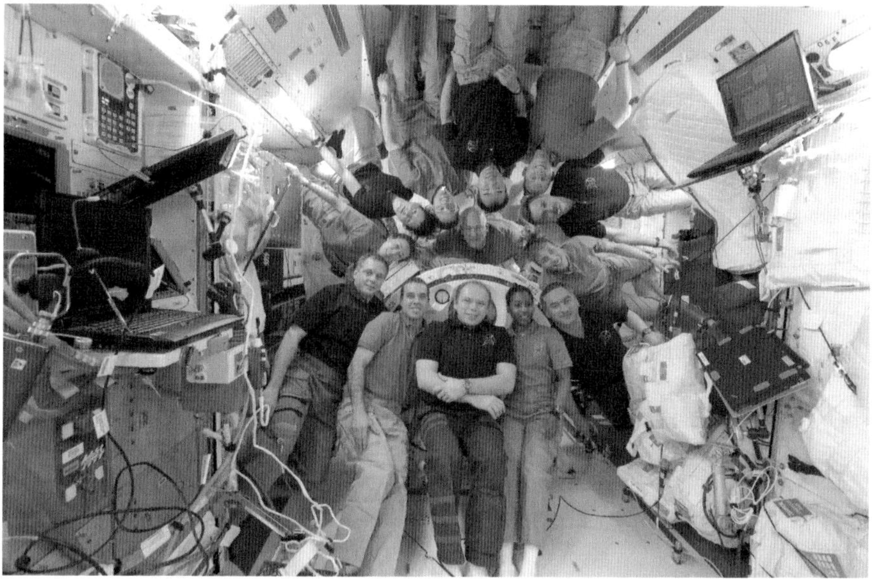

FIGURE 4.3 Astronauts onboard the ISS pose for a photograph. Some are oriented in line with the camera, while others are not. On the far left, a local vertical has been established: a workstation is set up with tools that are aligned with the orientation of the camera, as indicated by the orientation of the keyboard and the written labels on surrounding objects. © NASA

> trying to remember how the interior was oriented when you removed the equipment. This can continue for a long time, until you find the right position. (Battaglia 2012: 1090)

Judgments about the intended or natural orientation of an object are made both as a result of the characteristics of that object and its position within an "environmental field" (Dilworth 2005: 151). This environmental field, on Earth, is aligned with our culturally standard axes, and reference to this field informs judgments about whether an object is upright or upside-down. Dilworth (ibid.) recognizes several possible cues to the orientation of an object: the features of the object, and its position within the environmental field. An object that has faces that are identifiable as top, bottom, or other features that suggest an intended or natural orientation, can be considered to have "intrinsic orientation." Where such signifiers of orientation are absent, the observer can look to the environmental field for indications of the object's orientation, including relationships to other orientations, and infer information about the object's orientation from the surrounding evidence. This Dilworth describes as "field orientation." An object is "generally (perceived as being) *upright*... [when its] intrinsic top is *aligned with* the top of its environmental field." On Earth, therefore, "the orientation of an object is... a relative phenomenon" but it is not experienced as such (Arnheim 1960: 66).

Within Earth gravity, orientation is inextricably linked to stability. The tendency for sculptural objects to rest on a stable base evidences the extent to which sculptors take gravity for granted. Even works that do not have a stable base balance in ways that evidence engagement with balance, stability, or instability, as consequences of gravity. Lisa Pettibone identifies as an example Alexander Calder's *Untitled* (*c.* 1940), a sweeping wire and metal sculpture that rests at its midpoint. "Calder chose to demonstrate an object's pull to the earth by playing out its gravitational qualities in mid-air ... sitting between the two primary sculptural types he developed, namely mobiles (suspended from the ceiling) and stabiles (resting on the floor)" (Pettibone 2018: 13). Balance, like orientation, is a phenomenon that arises from gravitation. In microgravity, there is no correlation between orientation and balance, as an object is not destabilized by being tilted from its intrinsic orientation.

In the production of artwork, "artistic intentions or cultural norms play a primary role in determining which side of a [work] counts as the top" (Dilworth 2005: 151). Sculptor Arthur Woods (1993: 299) observes that, "consciously or unconsciously, artists conceive and [create] with gravity determining the eventual resting point of the work ... Our response to its aesthetic 'rightness' is based on our own experiences within the terrestrial environment." Those cultural norms are exhibited within the work itself or within its presentation: orientation may be determined by a frame, a plinth, the walls of a gallery or other objects in the environmental field (Arnheim 1960: 67). Artists may choose to adhere to or defy cultural norms in their work: to produce work that appears to be intrinsically oriented in line with the environmental field, or to produce work that is not in line with the culturally standard axes and may, therefore, appear to defy gravity. Consider, for example, Rebecca Horn's *Concert for Anarchy* (1990), a piano suspended from the ceiling by its legs, hanging "solidly yet precariously in mid-air" (Taylor 2003). Horn's piano, suspended counter to its intended orientation, is, nonetheless, visibly affected by gravity. The weight of the piano and the bending of its keys—both the result of gravity—demonstrate that it is hanging, not floating. We are so accustomed to encountering objects that are gravity-congruous, that when we encounter objects at unfamiliar angles, they are made strange. Unfamiliar orientations may, observes Arnheim (1960: 70), "bring to the fore a new structural skeleton that gives the object a different character." These alternative views may be interpreted as "deviations" from the expected or normal view of an object (ibid., 72). Horn's *Concert for Anarchy* may be aligned with in part "anarchy" because, on Earth's surface, an object that is upside-down is somehow incorrect or against the natural order; to upend an object is an act of rebellion.

In the context of spaceflight, upending is the natural order. Weightless objects do not necessarily appear upended, as in microgravity there is no upward or downward direction. Objects do not rest on a surface nor at a particular angle, and neighboring objects may have different orientations. A model commissioned by Peter Gutkin for the lobby of NASA's Ames Research Centre depicts three-dimensional objects as if they are weightless (see Figure 4.4). In this model, three objects are posed at different orientations. While each of these objects might be imagined to have a natural resting point on one of their faces, none are resting at what might be considered their "intrinsic orientation." In Earth gravity, for example, the cube on the far right of Figure 4.2 might be expected to rest on one of its six flat faces, but instead it is balanced on one of its vertices. Is it a result of their unusual orientation that these objects are understood as representing the behaviors that they might exhibit when weightless.

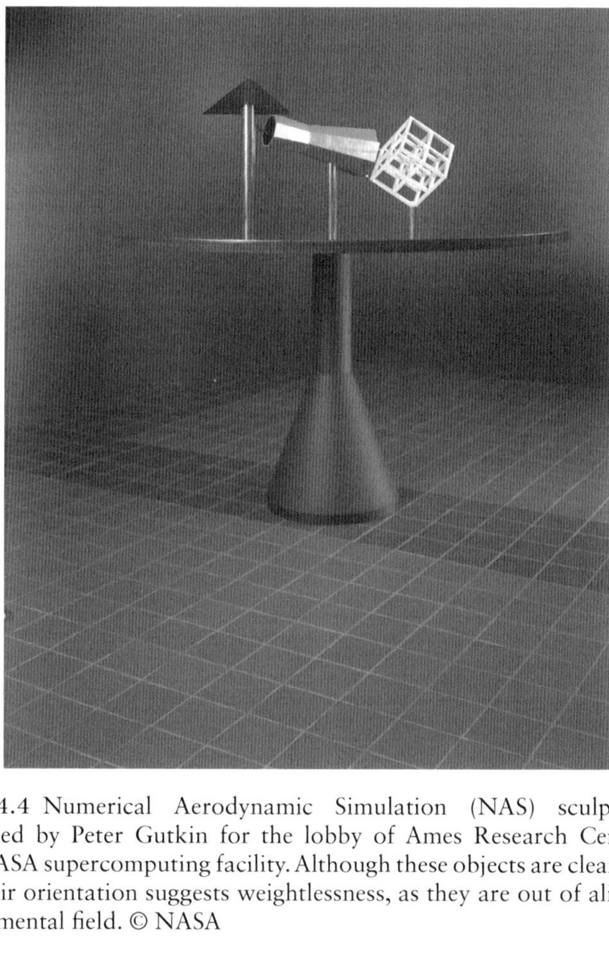

FIGURE 4.4 Numerical Aerodynamic Simulation (NAS) sculpture model, commissioned by Peter Gutkin for the lobby of Ames Research Centre building N-258, a NASA supercomputing facility. Although these objects are clearly supported by rods, their orientation suggests weightlessness, as they are out of alignment with the environmental field. © NASA

Objects oriented in unusual ways expose unfamiliar surfaces. This is the intention for Eduardo Kac in his *Space Poetry*, explorations of what Kac describes as "antigravitropism," beginning with his holographic work and, more recently, several works that have been deployed in space (see Case Study 6). Kac has employed the medium of holography to replicate the potential of the microgravity environment, where linguistic objects could be rendered in three dimensions and revolve so that they present alternative faces. Kac's "holopoems," including *Andromeda Souvenir* (1990) and *Astray in Deimos* (1992), invite audiences to move around the work so that they are exposed to alternative views of the three-dimensional linguistic forms that appear to float in holographic space. As a viewer moves, they observe different configurations and faces of the holographic forms. This causes letters to become abstract shapes, and vice versa. Kac (1997) tells us that these forms "escape the constancy of meaning a printed sign would have." Kac considers the orientation of his forms not only in relation to the audience but also in relation to one another. He developed an "orbital syntax" in which letters float around each other in a "poetic geostationary orbit" (Kac 2021: 81). Having been given the opportunity to engage directly with the environment of space onboard the ISS, Kac deployed these ideas in sculptural form, in *Télescope Intérieur* (*Inner Telescope*, 2017), a three-dimensional work that is intended to be viewed as letterforms from some angles, and a sculptural form when viewed at other angles. Drifting weightlessly onboard the ISS, the work revolves and presents these multiple identities.

Non-oriented encounters with objects occur not only due to the weightlessness of objects, but also due to the weightlessness of the person who encounters them. The object and the viewer may be differently oriented to one another. While it is true that any sculptural work may present a different face to viewers who are located in front of, behind, or beside it, the number of possible views is multiplied when the work and its audience are both weightless. Even an object that is stabilized at a particular orientation relative to its environmental field may be encountered by an astronaut whose own body is oriented differently. The orientation of a weightless object cannot, therefore, be considered in isolation. Both the object and the viewer must be considered. Kac (2007: 120) suggests that the weightlessness of the reader transforms his *Space Poems* into performances. In documentation for *Télescope Intérieur* Kac (2021: 18–20) presents sketches of his work in microgravity. These sketches depict the work itself, at multiple orientations, and, in addition, a human figure floating weightlessly around the work. Notably, the human figure is dominant in these images, being larger and often more centrally positioned on the page. The figure twists and turns, always with a gaze fixed on Kac's sculpture. The implication is that the weightlessness of the viewer is as essential to the experience of *Télescope Intérieur* as the weightlessness of the sculpture itself.

Just as the artefact is disoriented in space, so too is its audience. The artificial environment on Earth is designed around upright bodies. Tim Ingold (2004: 315) describes a "bias of head over heels" that shapes Earth-bound human existence, and our encounters with objects in our environment. The human body is "attuned to Earth gravity conditions" (Jörges and López-Moliner 2017). Proprioception tells us, even in the absence of visual cues, whether our bodies are vertical. We are able to sense whether we are in a gravity-congruent or gravity-incongruent position. This phenomenon is sensed physically through the body's vestibular system. Thus, even when movements are not aligned to an agreed vertical, there is a persistent sense of bodily orientation. The position of the body may contribute to the perception of an encountered object as being appropriately or inappropriately oriented. The vestibular system that is responsible for telling the body whether it is upright is disrupted in microgravity. Bodily cues that might indicate orientation or balance are absent. The body in space never feels upright or upside-down, rather, it falls as though it is endlessly tumbling without potential for resolution (see Chapter 5). The physical sensation of one orientation is indistinguishable from that of another.

Earth-bound humans are accustomed to performing most bodily movements along the plane of the Earth's surface, "restricted to the anatomical planes (sagittal, transversal, and coronal planes)" (Ahmadi 2020: 5). Microgravity introduces a range of movement possibilities that harness the constant state of freefall in which orbital bodies exist. Weightless bodies can "pitch, yaw, and roll" (ibid.). This enables bodies to inhabit the space within the cabin of a spacecraft as a fully three-dimensional space, in which locomotion can carry a space traveler in any direction, not only along the plane of a surface. Space travelers inhabiting a shared environment often do not share the same longitudinal axis, and so may not encounter each other at the same orientation. A space traveler may meet another passenger feet-first, or they might approach a fellow passenger as if from above, encountering the top of their head and shoulders before their face. In this way, human physical interactions are made strange by this environment. This multiplies the ways that audiences might interact with art objects. Multiple users might encounter an object from different directions and orientations at the same time. Frank Pietronigro (2004) observes that, in his *Drift Painting*, paint strokes are not confined to a single composition on a two-dimensional surface (see Chapter 2). Pietronigro deposited paint from within the inside of a three-dimensional canvas. For the artist and audience, there are "an infinite number of compositions happening simultaneously, each in accordance to all points of view from which the work is viewed." The position and orientation of each viewer determines the composition that they will encounter.

Those who have experienced weightlessness describe orientation in microgravity as a subjective phenomenon (see Chapter 5). Choreographer

Kitsou Dubois (2003) describes how, in microgravity, weightless bodies must build their own structures "around a subjective axis." Dubois has explored bodily movement onboard reduced-gravity aircraft as a dancer and choreographer, and describes how the weightless body must find "a new system of references" for self-orientation. She observes that the movement of the eyes locates objects in or out of alignment with "body's longitudinal axis, rather than the gravitational vertical" (Dubois 1994: 60). Here, the directionality of the body—the longitudinal axis (head to toe)—plays a more significant role in the perception of space than the body's orientation relative to a plane or surface within their environment (du Preez 2022: 1).

New approaches to orientation

Weightlessness necessitates new approaches to orientation that accommodate the possibility that work may be viewed from above, below, and any angle in between. An artist may choose to embrace the unpredictability of the viewer's orientation, or may impose orientation cues or constraints that ensure that the work will be viewed as the intended orientation. The spaceworks featured in this book, along with an informal survey of the interactions that astronauts have with objects in microgravity, as recorded in NASA video libraries, suggest that there are three possible approaches to approaching orientation in works produced for weightlessness (see Figure 4.5):

- A situated approach, where orientation is in line with a local vertical as indicated by cues in the surrounding environment;
- A subjective approach, where the object's orientation is in line with the user but not necessarily with the environment;
- A non-oriented approach, in which designed objects have no agreed top or bottom.

The situated approach acknowledges that, in the absence of vestibular cues to bodily orientation, space travelers may seek visual cues to orient themselves within their surroundings. In a situated approach, the terrestrial experience is replicated in a local environment, inasmuch as an object is aligned with other objects in the setting. This setting is the environmental field that can be used as a reference point for judgments about the orientation of objects within it. Visual cues indicate that there is an intended orientation for any object within that setting, prompting users to orient themselves in line with the objects. This approach can be seen onboard the ISS, where workstations are set up so that tools are all attached at the same orientation, and the astronauts' work envelope can accommodate the necessary tools for a task when they are at an agreed location and orientation.

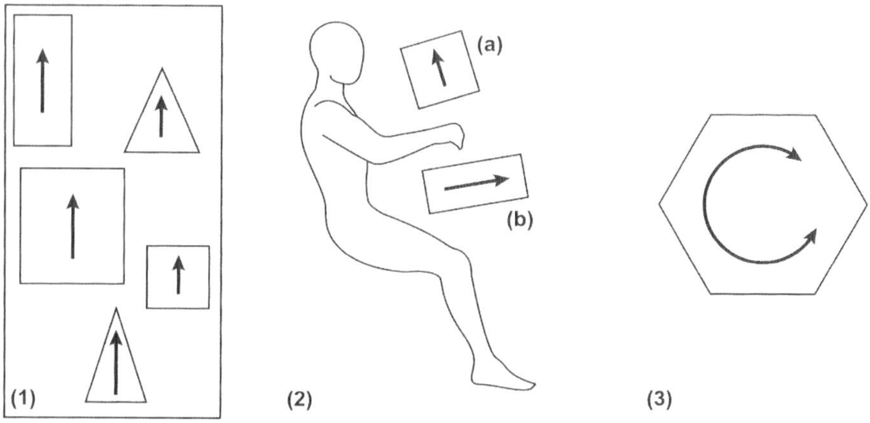

FIGURE 4.5 Three alternative approaches to orientation: 1. A situated approach, where orientation is in line with a local vertical; 2. A subjective approach, where objects may be orientated to suit the user, and could be a. aligned with the body's vertical, or b. some other orientation relative to the user, for example, oriented laterally to the body; 3. A non-oriented approach, where objects do not have a designated top or bottom and could be used at any or multiple orientations.

The situated approach prompts a compartmentalized view of spaceship interiors, where activities take place within local, designated workspaces. Where two-dimensional works are displayed onboard the ISS, it is commonly in line with a local vertical. In Figure 4.3, in the far left of the image, an ISS workstation has been set up with a laptop and surrounding tools at the same orientation. At this workstation, a "local vertical" has been established, as objects required for a common task have been aligned with one another in a local environmental field (Bell 1983: 3). This way of conceiving workspace is common throughout the ISS, following conventions established in the "individual experiment setups" onboard Skylab, and now seen in the exchangeable modular racks onboard the ISS, which is itself modular (Sherwood 2010: 1–2). Bell (1983: 3) describes methods that can be used onboard a space station to prescribe a local vertical using "colors, lighting, and position of dominant focal elements." These are sometimes supplemented with functional décor that acts as an orientation cue (Coss et al. 1989).

The situated approach is suitable for any work that was intended for the terrestrial environment, or that is intended to be viewed in a way that resembles how it may be encountered on Earth. This approach attempts to erase the effects of weightlessness, replicating as closely as possible how an artefact may be encountered in a setting where there is a stable paradigm of orientation. Where professional astronauts and cosmonauts have affixed their own private collections of artwork to the interiors of the ISS, they have tended to take a situated approach. Photographic records of the Zvezda service module in 2001 show landscapes, religious icons, and astroart variously attached and strapped to a cabin wall, all at the same orientation. Although a crew workstation and laptop are shown at a different orientation, the paintings and prints collectively contribute to an informal gallery that establishes its own local vertical.

Where objects are not aligned with a local vertical, or are not affixed to the interior of the spaceship or space station and can be freely rotated by the user, a subjective approach to orientation can be taken. Such subjective approaches to orientation may be seen where there is an intention for an artefact to be manually handled, or oriented for observation by only one viewer. Where orientation is considered subjectively, the user's body dictates the most appropriate orientation for the object. Dilworth (2005: 151) employs the term "observer-relative" to describe such orientation. This approach is familiar on Earth where we take a subjective approach to the orientation of hand tools, in which they are operated as extensions and augmentations of the hand (Buckminster 1965: 61; Holmes et al. 2004). As such, we tend to consider their orientation relative to the body, rather than relative to a stable paradigm of orientation, so that they are typically oriented outward rather than upward—with ends designated as *proximal* (inward-reaching; handle) and *distal* (outward-reaching; tip), as opposed to top and bottom. The microgravity environment creates the possibility of other objects, potentially larger and more unwieldy than hand tools, being designed for subjective orientation. In these cases too, it can be helpful to consider anatomical direction, where the absolute values, *top* and *bottom,* are replaced with subjective values, *superior* and *inferior.*

Wearable artefacts are perhaps those that are most explicitly subjectively oriented. Shachar Freddy Kislev's *Space Jewel* is one such example, a ring that houses metal "follicles" in a row of pores (see Figure 4.6). The ring was transported to space on the Rakia mission in 2022 and worn by an astronaut in the microgravity environment of the ISS. When weightless, the follicles float away from the ring, remaining tethered by short chains. They float within a movement envelope that is determined by the length of the chains and the position of the wearer's hand. Efforts were made to capture the ring's anticipated behavior prior to its launch into space. Photographs record the ring rotated upside-down, so that the follicles hang downward. In this image the chains are pulled straight, perpendicular to the finger, by the weight of the follicles and the force of gravity. Digital simulations made before the ring was launched into space revealed that turning it upside-down could not accurately represent how it might behave when weightless. The digital model demonstrated that the follicles and their chains would not extend directly outward, but rather, cluster and then separate, each appearing to move independently and aimlessly through space. This behavior is confirmed in footage captured onboard the ISS, where the follicles are seen "dancing" and floating as if underwater (Noble 2022: 68).

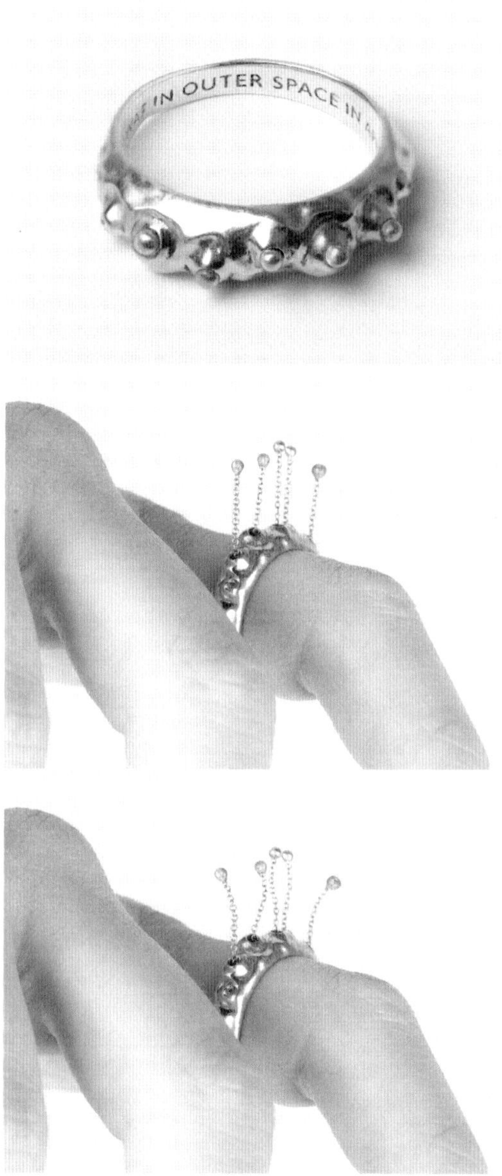

FIGURE 4.6 Shachar Freddy Kislev, *Space Jewel*, 2022. Three views of *Space Jewel* show Kislev's work in Earth gravity and microgravity. Metal "follicles" float away from the wearer, tethered by chains to the ring. In the second image (middle), the ring had been turned upside-down, and the photograph rotated, in an effort to show how the ring might behave when weightless. Here, the chains are pulled taught by the force of gravity, perpendicular to the finger. However, simulations and videos of the ring in microgravity reveal that the behavior of the chains and follicles is more dynamic. The chains move more fluidly, and the follicles cluster, separate, and dance around the ring (bottom). Photo: Achikam Ben Yosef, courtesy Shachar Freddy Kislev.

Potential for both a situated and subjective approach to orientation exists in the virtual environment. Augmented reality (AR) devices are typically worn or hand-held, and so, when they overlay virtual objects onto an environment, these objects are positioned relative to both the environment and the user. AR devices function by tracking orientation and location in an environment, and overlaying virtual objects relative to that data. Logan Smith (2022) outlines how AR can detect high-contrast visual cues in an environment to detect the orientation and location of a device, thus enabling AR devices to operate in microgravity. Differently from how AR devices work on Earth, where accelerometers and magnetic field sensors detect which way is up, in microgravity the orientation of the device is one way of determining a subjective upright, and this can be tracked using visual cues in the environment. Augmented reality is used onboard the ISS to help astronauts complete tasks more efficiently, by overlaying diagrams or schematics onto astronauts' views of real objects (Gaskill 2021). There is potential for art applications of this technology, to locate virtual art objects in the environment of space, or to extend the environment of the cabin of a spacecraft into virtual space. These possibilities have begun to be explored by Martin Sjardijn, whose *Black Box* (2018), as a result of its virtual existence, is without weight or mass, but can appear in any location in space. Existing only in augmented reality, its size, location, and orientation are always relative to the view, and, by scaling and reorienting this virtual object within the image plane, it can appear as a small object onboard a spaceship, always reorienting as the viewer moves, or a large object of cosmic proportions in outer space.

The third approach—a non-oriented approach—is perhaps the furthest from the familiar approaches that designers may be accustomed to taking in Earth gravity, but, arguably, this is the approach that is most natural for the weightless environment insofar as it follows the lack of a fixed paradigm of orientation. The non-oriented approach may involve thinking in the round, where there is no requirement or intention for the object to be oriented in any particular direction. Consider the example of the Earth itself, a spherical object with no top or bottom except for that arbitrarily designated by humankind. The sphere is a shape that many objects and materials adopt in space, shaped by their own gravity. Numerous spaceworks adopt a spherical shape, including Peter Beck's *Humanity Star* (see Figure 1.1), and Kirsten Johannsen's *Nomadic Nature Kit* (2010, see Figure 4.7). *Nomadic Nature Kit* is a microclimate sealed in two domes that connect to create a sphere. Johannsen intends the work to function as a personal, portable garden that astronauts can carry with them during space travel, and that will float freely through the interior of a spacecraft. Johannsen's work seeks to help space travelers overcome a sense of "placelessness" that they may experience during spaceflight (Johannsen 2010). The work combines "'here' and 'there'": the plants provide a reminder of Earth, and yet their direction of growth is a reminder of the unearthly environment in which the spherical garden is encountered.

The intention of a spherical spacework may be to present an identical or comparable view regardless of how it is encountered or oriented. This is not necessarily the case for every non-oriented spacework. The artist might intend for different views of their work to present different identities, as in Kac's *Télescope Intérieur*, where no orientation is correct or definitive but, rather, each angle reveals a different meaning. These meanings may be revealed over time, as the object revolves and the viewer navigates around it, accessing additional visual information as they move. A non-oriented work may, therefore, be understood as a temporal experience, created with the intention that its audience will experience it at a variety of angles and orientations. When Arthur Woods produced *Cosmic Dancer* (see Figure 4.8), his intention was to create a sculpture that has no resting point, and "no natural 'top or bottom' orientation" (Woods 2019). The geometric,

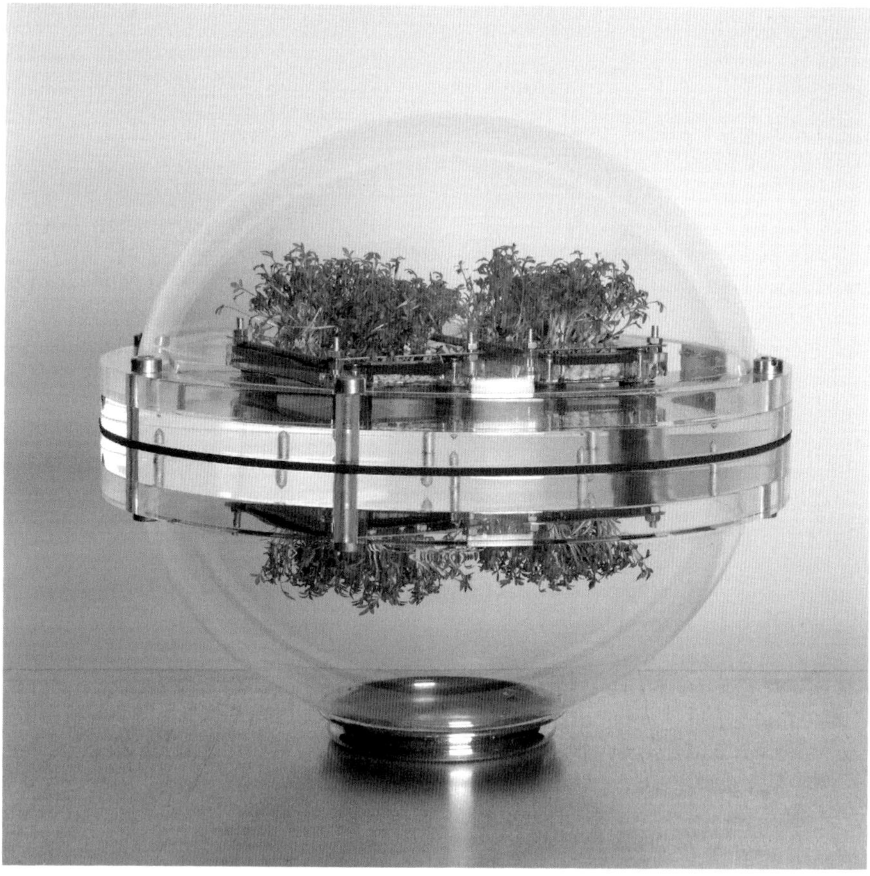

FIGURE 4.7 Kirsten Johannsen, *Nomadic Nature Kit: Artwork for Long Term Space Travellers*, mixed media, object assembled 40 × 40 × 38 cm, 2010. © Kirsten Johannsen + VG Bild-Kunst, 2010. Image © Heinrich Hermes, courtesy Kirsten Johannsen

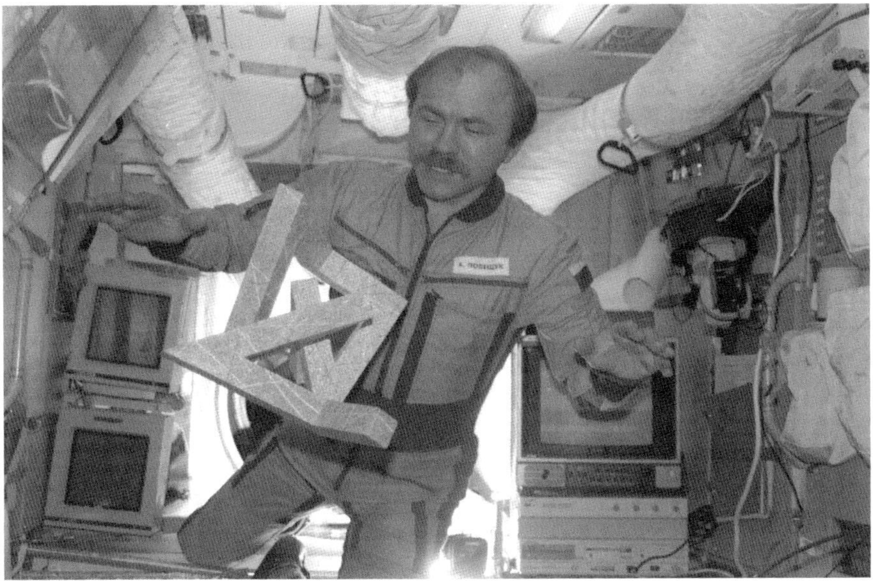

FIGURE 4.8 Arthur Woods, *Cosmic Dancer,* acrylic, aluminium, 35 × 35 × 40 cm, 1993. *Cosmic Dancer* is shown here onboard the Mir space station, with Cosmonaut Alexander Polischuk. The sculpture has no intended resting point or orientation, and in this way is reflective of the lack of an agreed paradigm of orientation in the microgravity environment. Image courtesy Arthur Woods.

painted aluminum sculpture was launched to the Mir space station in 1993, where cosmonauts interacted with the work and can be seen, in video documentation, nudging its corners to adjust its angle and speed of rotation.

The non-orientation of Woods' *Cosmic Dancer* is highlighted as a result of this work being one in a series. Each sculpture in the series is laid so that it rests at a different orientation. As a result of its resting point, each sculpture appears to take on a different character, including, for example, *Reclining Dancer* (1993), which, through the anthropomorphism suggested in its title, makes references to the ways in which the human body experiences a range of strains and pressures in different postures and positions. A larger sculpture, also called *Cosmic Dancer* (1985) is installed at the Specogna Complex in Switzerland, where it rests at a fixed orientation. Resting so that one of its limbs lies along the ground, it appears balanced and heavy, as if this is the object's most natural or intended orientation. Woods (2005) identified several possible resting points for these works, but allowed gravity to be the final determining factor in their orientation. When the smaller version of Woods' sculpture was released in a microgravity environment onboard Mir, the sculpture fulfilled its potential to be "viewed from any perspective and angle" as it floated freely in microgravity. While weightless, Woods' sculpture can be seen "in a state of perpetual change," in that its orientation, and hence composition, is continuously altered as "playfully dances" through the astronauts' habitat (Woods 2021a).

Eduardo Kac (2005: 21) views engagements with disoriented space as a natural progression from twentieth-century developments in sculpture that saw artists including Alexander Calder "liberate modern sculpture from the constraints of enclosed and static form resting on the two-dimensional surface of the pedestal." As with these previous developments, the new environment allows artists to reconceive space and form. Weightless objects are, suggests Kac, "truly three-dimensional" in a way that is only possible in microgravity. It is only possible, in this environment, for an object to present all of its faces: to be encountered at any orientation, and for all of those views to be visible simultaneously by different observers. Some artists view this as a progression from the idea of a sculpture of a three-dimensional object to the sculpture as a four-dimensional object, thus taking the non-oriented approach a step further. Mario Ramiro (2000: 104) imagines a future in which "far away from gravity, we will be able to design and assemble gigantic frameworks without any concern for the classical notions of above and below," where "sculpture will acquire new dimensions sprung from new structural relations."

Progress toward this future is found in Arti Struyanskiy's submission for the *Moon Gallery*, for which the artists took a non-oriented approach that imagines that orientation can be transcended. *Human Frontiers* (2021, Figure 4.9), is a thread and mesh sculpture that invites viewers to perceive "all sides of a structure at once." Struyanskiy's use of colored polyester mesh in this and his previous sculptural work, *A New Era of Thought 1888–2020* (2020), allows viewers to look through a cube to see all faces simultaneously, blurring the divide between time and space. The structure is held together by threads that loosely resemble the edges of a cube but are woven through the shape. As a result of Struyanskiy's use of flexible materials, the structure is held together more by tensegrity than the rigidity of any of its parts. Despite the sculpture's small size, Struyanskiy hopes that the experience of *Human Frontiers* will be of an object that is limitless. He selected six different colors of mesh, so that as the "colors become a blended visible spectrum where one color no longer can be separated from the rest," "moving around the structure endlessly transforms [its] appearance" and the object appears to be continually evolving.

These three approaches to orientation are not mutually exclusive, and not necessarily dictated by the artist. A non-oriented object may become oriented if fixed to a surface. Similarly, an object that has been aligned with a local vertical may be detached from that setting and rotated to match the longitudinal axis of a body that is oriented differently. As an artefact's orientation is always relative, it is the product of its environment and its audience as well as the properties of the artefact itself. At this stage in the evolution of orbital art, many spaceworks are accompanied by explicit instructions that itemize the ways in which astronauts should experience and

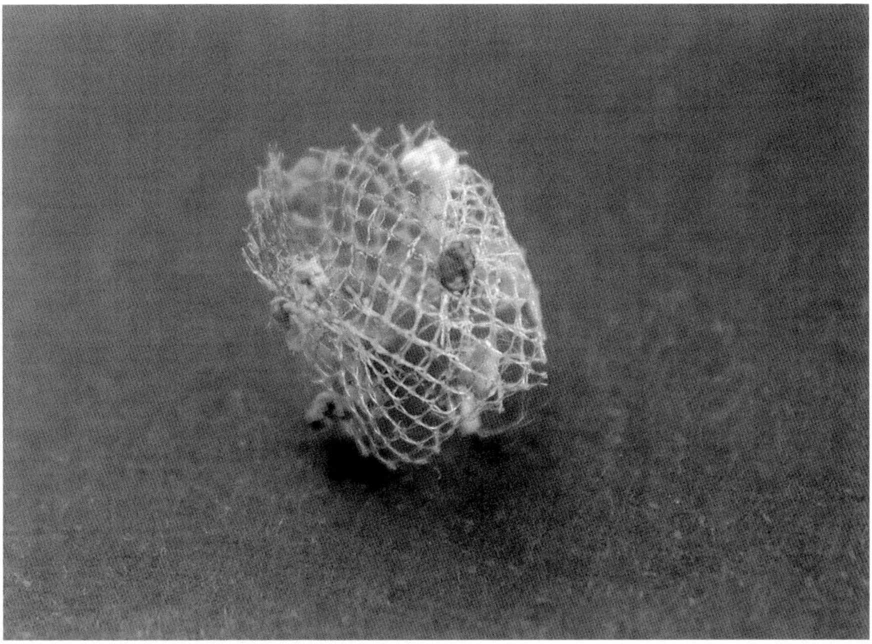

FIGURE 4.9 Arti Struyanskiy, *Human Frontiers*, polyester thread and mesh, 1.5 × 1.5 cm, 2021. Struyanskiy explores how sculpture can represent four-dimensional space-time, in which all surfaces of an object can be viewed simultaneously. *Human Frontiers* was launched to space as part of the *Moon Gallery* in 2022. Image courtesy Arti Struyanskiy.

interact with a work (see Chapter 2). As a result, artists are able to dictate the way that a work is deployed and communicate expectations about how viewers may interact with it. Although this is not peculiar to space—galleries on Earth also sometimes dictate the situation in which audiences encounter a work—it does limit the potential for a work to transition from one approach to orientation to another without explicit direction from the artist. In the future, space travelers may become curators of collections of artefacts and make unexpected decisions about how those artefacts are exhibited within their habitat, and it is possible that those habitats may be designed to accommodate orientation differently. New curatorial practices may emerge that are specific to the weightless environment. In these new contexts, spaceworks will not be considered in isolation. Decisions about the orientation of an individual artefact may be informed by the nature of a larger collection of spaceworks.

Case Study 5
Eyal Gever, *#Laugh*, 2017

Eyal Gever's *#Laugh* is a three-dimensional representation of a human laugh, printed onboard the ISS in 2017 (see Figure 4.10). The work was generated from a spectrogram of a human laugh, looped into the shape of a ring. The circular shape suggests an endlessly looped sound, as infinite as the void of space. The work does not have an intended orientation: the soundwave from which it was generated is more recognizable from what might be considered the lateral view, while the loop is more visible when rotated 90 degrees (see Figure 4.11). Meaning, therefore, is revealed as the work revolves through space. The size of the work invites manual interaction and, consequently, invites reorientation. While the sculpture may freely rotate as it drifts through the ISS, astronaut viewers may grasp it in their hands and choose to inspect and reorient it, in search of a particular view, or in search of meaning. The *#Laugh* is both personal and universal,

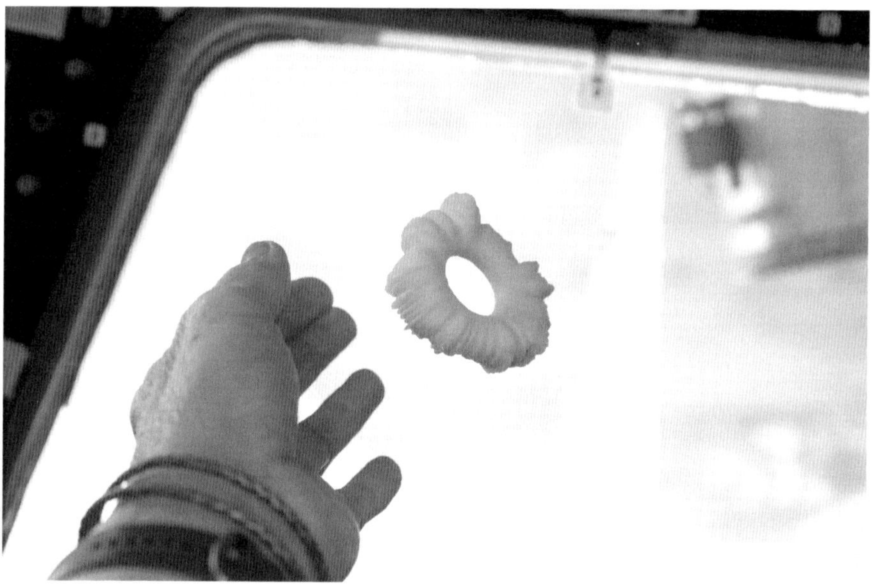

FIGURE 4.10 Eyal Gever, *#Laugh*, 10 × 10 cm, 2017. Gever's sculpture is shown here being directed by the hand of an astronaut onboard the ISS. The size of the object invites manual interaction, which leads to the stabilization of the work, at an orientation that is fixed relative to the body of the astronaut. This size was dictated, to some extent, by the technical requirements of the project, including the size of the printer bed, and the amount of electricity and time that Gever was permitted to consume during printing. The white color was also dictated by the materials onboard the ISS. Gever had originally intended a rainbow-colored object, with a color scale that represented the varying amplitude of the recorded laugh. © NASA/Eyal Gever

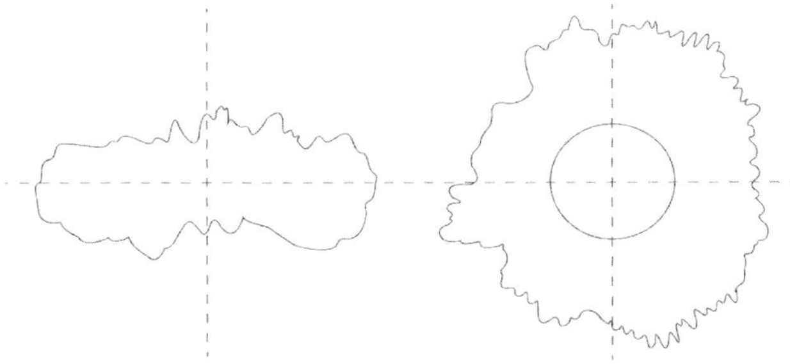

FIGURE 4.11 Diagram showing two views of *#Laugh*. The spectrogram view (left) is more clearly derived from the visualization of a recorded laugh. The ring view (right) reveals how the soundwave has been looped, and prompts comparisons with a star. Other views are also possible as the sculpture floats weightlessly through space.

a record of an intimate expression of emotion by an individual, as well as a universally familiar "sound of humanity."

#Laugh is the result of a collaboration with Made In Space, a commercial provider of 3D printers for space. The printing of the sculpture onboard the ISS meant that it did not have to be transported to space from Earth, but rather could be produced using equipment and materials that were already located on the ISS. Jason Dunn, co-founder of Made In Space, suggests that one of the main barriers preventing progress in space exploration has been dependence on Earth, including the Earth-based manufacture of objects for spaceflight. The potential to 3D-print objects in space significantly advances opportunities for engagement with this new environment, increasing the efficiency of missions by enabling print-on-demand as opposed to a requirement to transport heavy payloads from Earth, containing tools to account for every possible eventuality. This efficiency benefits would-be space artists by decreasing the potential cost of their projects. This new opportunity for artists arises only, however, when those who use technologies such as the 3D printer recognize its cultural as well as functional value. Gever was given the opportunity to create work for space only because the founder of Made In Space has advocated for creative use of his technology.

The form of the *#Laugh* object resulted from crowd-sourced data and choices made by users of an app that was produced for the project. Users were asked to record eight-second samples of laughter and apply them to their choice of five geometric shapes. Users then nominated their favorite of the completed shapes, and the winner became the final sculpture. Gever initially collaborated with a large team of engineers at Autodesk in the creation of an earlier version of the app. Six months prior to NASA's deadline, Gever abandoned this approach, concluding that the work so far had become too focused on the tool itself, and not on the art. He cut his team down to three engineers and rewrote the app, this time focused on creating an "artistic experience" that would immerse his users in the journey of creation.

Gever views it as his responsibility to understand the technology that is employed to produce his work. He describes his role as being "capable of understanding cutting edge technology and being expressive with it." For Gever (2016), "art needs to be [a reflection] of the time that the artist is alive as well as to push the art language forward [*sic*]" and, for Gever, this means engaging in the forefront of human technology, as well as the new environments that humans inhabit as we become a spacefaring civilization. He reflects that, as an artist who often makes use of cutting-edge technology, it is important not to become "overly excited by new features, and remember that [the purpose of] an artwork [is to] touch people"; that "technology [is] a means to an end, but the end is the art."

For Gever the biggest challenge was not the technicalities of producing work for microgravity, but the question of how the work could "touch people's souls without creating controversy." It needed to be "genuine, and have an international appeal, interacting with politics, with culture, with history." At the suggestion of a friend, he chose laughter as the sound that could be "totally universal," transcending language and culture. If the work were to represent all of humanity, then it was important for Gever that consensus be sought as to how the crowd wanted their species to be represented. The laugh, and the decision about which shape would become the final sculpture, needed to be the result of "the wisdom of the crowd." He cites as inspiration prehistoric hand paintings that were produced by individual hands but have come to represent a whole society and evolutionary era. Those hand paintings asserted, "We are here," and, with *#Laugh*, Gever set out to send a similar message out into the cosmos. Comparisons can also be made to the Golden Record, a phonograph record launched into space onboard the Voyager spacecraft in 1977, which featured an Earth soundscape and greetings spoken in multiple languages. Much like the Voyager record, *#Laugh* encodes the sound of humanity in a physical object that has the potential to be played as an audio recording. There is potential for the sound of the laugh to be reproduced by some future or "other intelligence."

Having relied on the mechanical technology for the manufacture of his work, Gever was pleased to see an imperfection in the printed object, arising from the need to print the object in two parts. He describes the appeal of the imperfections that are visible in humanmade art, such as painting, as evidence of the artists' human touch. "Surprises and mistakes," and evidence of "limitations," he suggests, make art human. The "tiny gap" between the two halves of the *#Laugh* object are evidence of human problem-solving, and the current limitations of humanmade technology.

As the *#Laugh* object was printed onboard the ISS, Gever has not seen it directly or held it in his hands. Further, there are no plans to send the work to Earth. It exists, for Gever and most of his audience, only through documentation. In the development phase of the project, Gever produced

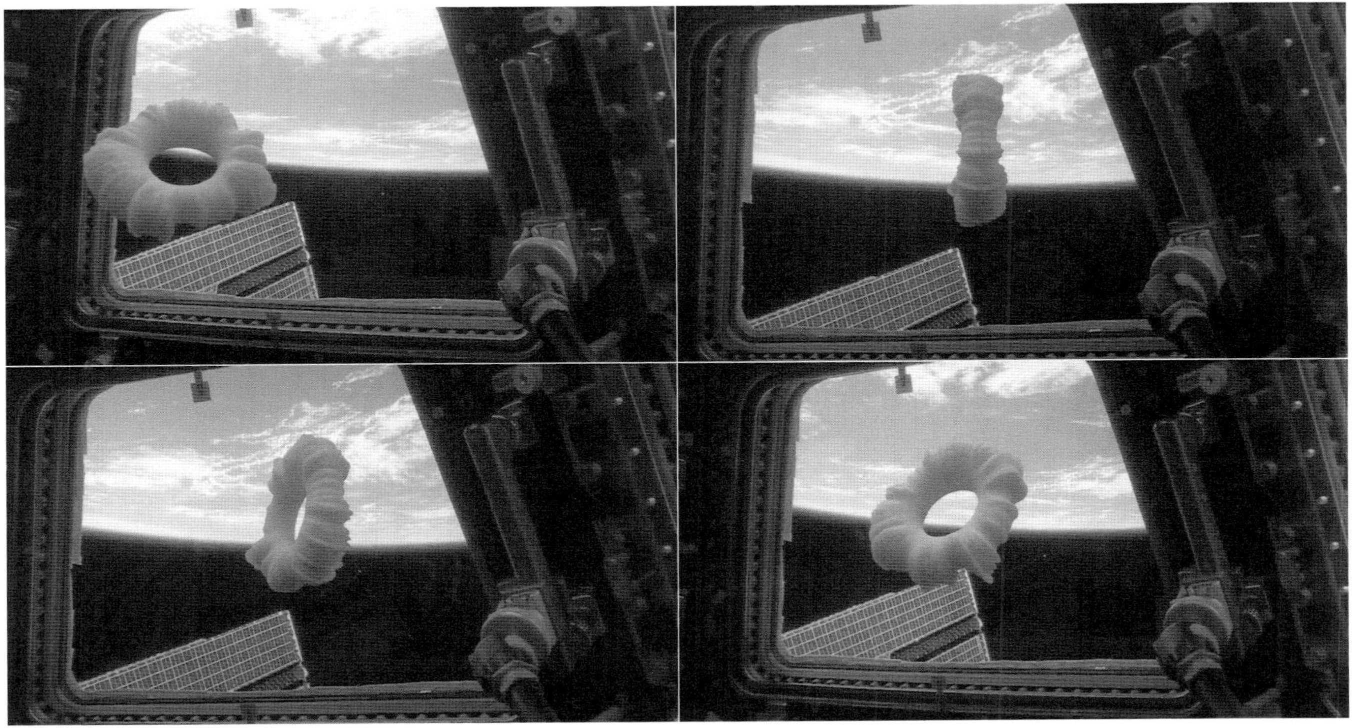

FIGURE 4.12 Video stills showing #*Laugh* floating inside the ISS observatory. As the sculpture floats, it revolves to present its different faces. © NASA/Eyal Gever

storyboards that predicted the manual interactions that astronauts might have with his spacework. The storyboards depict the work in the context of the ISS, shown against a backdrop of the distant Earth, and when he was first shown visual records of the deployment onboard the ISS, Gever was overjoyed to see how closely they resembled the illustrations he had drawn. These storyboards are of particular importance to Gever as the only physical record of his development of the work.

Gever plans future engagements with the environment of space in the form of a work that will be transported to the Moon. This next work is likely to be more somber in tone. Since he started working on *#Laugh*, Gever has seen significant changes in the world's political, cultural, and environmental landscape. He began the project in a period of optimism, and the laugh recording reflects that positive record of humanity. He is approaching his next project with mixed views of the future of humankind. He knows that his next work will be made possible by the brilliance of scientists and engineers working in his home country, but that exciting developments in spaceflight are taking place against the backdrop of an uncertain future for humanity and our planet. While *#Laugh* signified a united humanity, the next work will urge caution and conscience. His reflections are a reminder that the artist's role in space is not just to demonstrate potential for creative engagement with this new physical environment, but also to represent the cultural contexts in which that work is created.

Eyal Gever was interviewed on February 13, 2023.

Case Study 6
Eduardo Kac, *Télescope Intérieur* (2017) and *Ágora* (1986–2024)

Eduardo Kac has produced and deployed work "from space," "to space," and "in space." In *Spacescapes* (1989), an alternating sequence of satellite and microscopic images is broadcast to slow-scan television sets; *Monogram* (1996), a linear illustration of Kac's signature, drawn to resemble the flight of a spaceship toward the Moon, was recorded on a DVD and sent to Saturn onboard the Cassini spacecraft in 1997, and orbited Saturn from 2007 until 2017. At the time of interview, Kac is focused on the fourth phase of the deployment of *Adsum* (2019–present), a cube-shaped sculpture that has been deployed in the ISS and will, in the project's fifth and final phase, be deposited on the lunar surface. Kac's extensive body of space work engages, in particular, with weightlessness and the multiple and non-oriented views that are possible when an artefact is encountered in this environment.

Kac explores the view from above in his ongoing series of installations, *Lagoogleglyphs* (2009–present). These large-scale, low-resolution logograms are positioned on Earth so that they may be captured in satellite images and viewed via Google Earth. The works are designed to be seen digitally, drawn at a resolution that has increased with each new installation, as "the size of [the] pixels [matches] the resolution of the satellite used to create the work." Although the satellite photographs of the work are captured from orbit, the work's intended audience is Earth-bound. Kac intends for the work to inspire "cosmic empathy," "the ability to transport yourself to space and think from that perspective" using digital tools such as Google Earth and the system of satellites that feeds it. It is possible to virtually navigate the globe via Google Maps, zooming in and out at each of the locations where the first five *Lagoogleglyphs* are sited. Kac's eventual ambition for the project is to "make as many as possible, so that I can rubber stamp the globe with lagoglyphic messages."

Kac's vision for "Space Art" and "Space Poetry" long precedes his actual engagement with space. Since 1983, Kac (2007: 120–1) has conceived of work that is, as he describes it, "antigravitropic." He identifies "gravitropism" in "all forms and behaviors created on Earth," including poetry, which is constrained by gravity and a human tendency toward "gravitropic" working. The conventions of printed poetry dictate top-to-bottom reading, following "an ordinary perception of reality, which is regulated by the action of gravity upon elements" (Kac 1991). Kac sought to develop "space where the linguistic ordering factor of surfaces is disregarded in favor of an irregular fluctuation of signs that can never be grasped at once by the reader." Initially, Kac explored these ideas through "holopoetry," which is "written in light and therefore weightless." In works including *Holo/Olho* (1983), *Abracadabra* (1984–5) and *Chaos* (1986), letters become intangible, floating objects, freed from the plane of the page on which written or typed poetry is more commonly found, presenting themselves to the viewer at all angles. Letters become spatiotemporal events that viewers can encounter at any angle, from any angle. "By writing with light," says Kac, "I was freeing myself from the constraints of not only the physicality of writing where ink penetrates, by gravity, the fibres of paper, but [also] freeing myself from ... the logic of writing top-to-bottom."

With the expansion of the spaceflight sector, Kac has had opportunities to engage with "the material reality" of weightlessness in space, that enables the production of "a genuine space poetry." *Télescope Intérieur* (*Inner Telescope*), a three-dimensional work that was deployed onboard the ISS in 2017, presents different identities depending on the angle of encounter. The work appears, when viewed from some angles, to adopt the shape of the letters "MOI" ("me"), but as it drifts weightlessly it revolves, loses its linguistic identity and instead resembles the form of "human figure with its umbilical cord cut" (see Figure 4.13). The letters of MOI are revealed as the sculpture revolves, with the form of the "I" perpendicular to the plane of the "M." Kac (2014) considers neither of these perspectives to be the definitive view, nor does he consider the sculpture to have an innate orientation, stressing that it "has no top or bottom, no front or back." Its path and rotation, as it drifts through the ISS, are not only the result of its weightlessness but also the airflow caused by the station's ventilation systems. *Télescope Intérieur* harnesses the "pushing and pulling" of this airflow to enhance the movement of

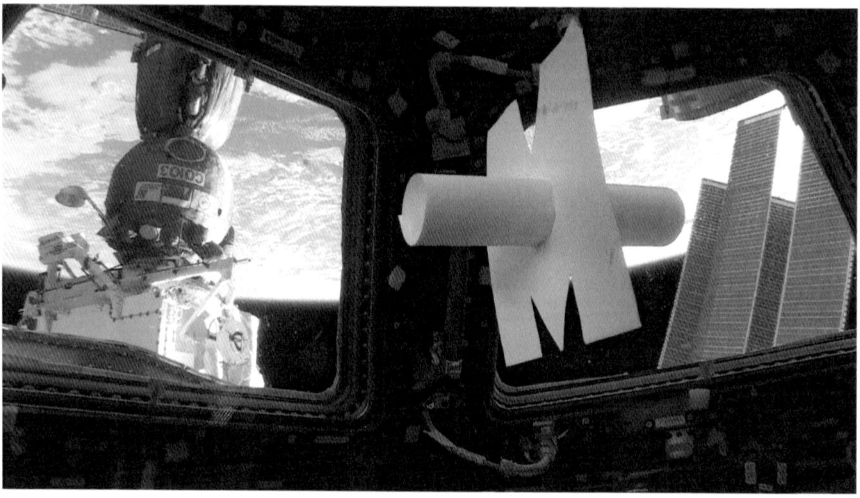

FIGURE 4.13 Eduardo Kac, *Télescope Intérieur* (*Inner Telescope*), space artwork, dimensions variable, 2017. As seen in the ISS, orbiting around the Earth. *Télescope Intérieur* was produced by the Observatoire de l'Espace (Space Observatory), the cultural lab of the CNES, with the support of the ESA. Image courtesy Eduardo Kac.

the work through the cabin, including the speed and extent to which it revolves and presents its different identities.

The nature of *Télescope Intérieur* is substantially shaped by constraints. As a resident of the Observatoire de l'Espace, Kac had an opportunity to deploy work onboard the ISS. The artist decided not to send any actual object to space, in order *de facto* to produce the work in space using materials and tools that were already onboard. Kac chose to work with paper and scissors, designing a three-dimensional object that could be constructed through a step-by-step process of folding, cutting, rolling and combining two sheets of paper to create two intersecting forms (see Figure 4.14). The first form is a flat plane—the shape of an M—formed by folding and cutting, and the second is a cylinder, formed by rolling, that is threaded through a circular hole in the center of the first piece. The work that Kac transmitted to space was a set of instructions—an entirely visual protocol. Each step of the construction of the paper artefact is illustrated in a sequence of line drawings with no text annotations (see Figure 4.15). These illustrations are an artwork in themselves, and have been exhibited in a vinyl installation and published as an artist's book that serves as a manual for potential further deployments of the work. Kac worked closely with the astronaut who would deploy his work in space, Thomas Pesquet. Together they rehearsed the process of construction, guided by the protocol. Each time, Pesquet attempted to replicate Kac's example, repeating the process until the result met Kac's approval. Several months later, Pesquet repeated the procedure in the microgravity environment of the ISS and transmitted photographs and images back to Earth as evidence of the activity.

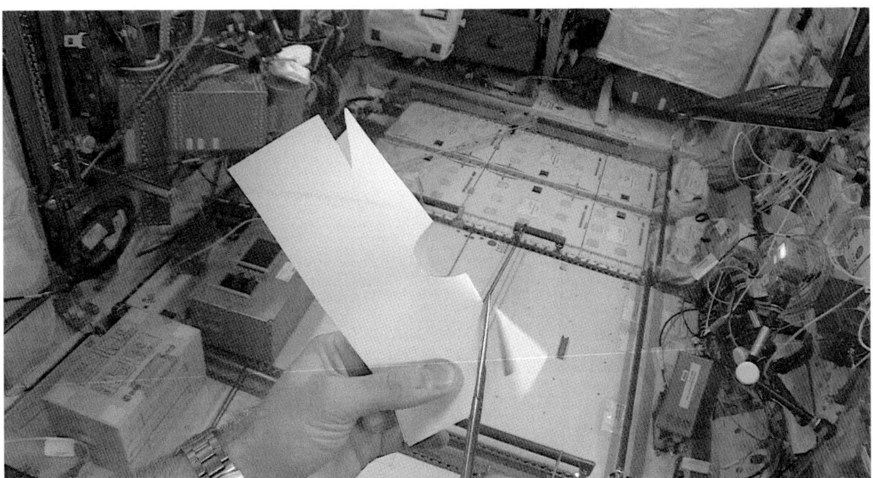

FIGURE 4.14 Eduardo Kac, *Télescope Intérieur* under construction in the ISS by astronaut Thomas Pesquet. *Inner Telescope* was produced by the Observatoire de l'Espace (Space Observatory), the cultural lab of the CNES, with the support of the ESA. Image courtesy Eduardo Kac.

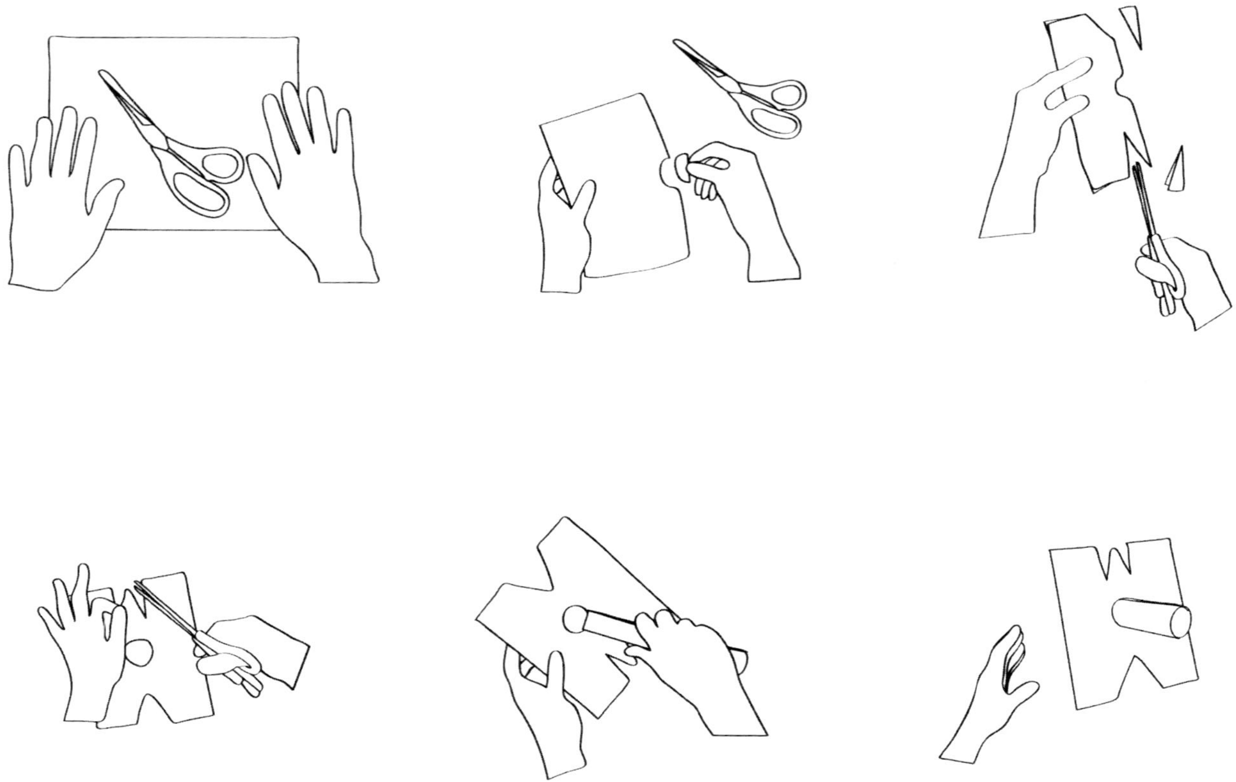

FIGURE 4.15 Eduardo Kac, excerpts from *the Télescope Intérieur* protocol, 2017. Image courtesy Eduardo Kac.

Kac resists comparisons to process art, neither viewing the astronaut as a collaborator nor as a performer. Kac (2021: 89) stresses that the work does not "privilege attitude and action over the physicality of the object" as one might expect with process art. He favors the term "protocol" over "event score," because his intention is for the work to be deployed exactly as per his intention, as opposed to allowing improvisation, as one might expect from a collaborator or performer. The goal is always, he stresses, the deployment of the object, not the performance of its production. It is Kac's intention for the role of the astronaut to change once the work has been deployed. The astronaut plays two distinct roles: "First, he takes the role of the astronaut that is executing the protocol, but once he finishes doing that then his role changes. Now he is audience." As audience, the astronaut is "our collective avatar," experiencing the work in the environment of space on behalf of those Earth-bound audiences who may never be able to directly encounter the work in space. Kac's own illustrations highlight the importance of the presence of an audience member floating weightlessly alongside the work. In series of drawings and embroideries, Kac depicts human figures at various angles and in an array of postures only possible in microgravity, illustrating that the non-oriented experience of his work is as much about the weightlessness of the audience's own bodies as it is about the weightlessness of the work itself, where both contribute to the multitude of angles at which the work is encountered.

Some of Kac's spaceworks have been decades in the making, conceived during his exploration of antigravitropism in holopoetry, and realized many years later when opportunities for deployment in space have finally arisen. *Ágora* was created for space in 1986, and finally launched in January 2024. As a young artist in Rio de Janeiro, Kac did not initially have the means or opportunity to deploy *Ágora* in space but, nonetheless, he recognized that the work could not be considered complete until it achieved this destiny. In the interim thirty-eight years, Kac chose not to exhibit the work, feeling that the idea of the work would not be transformative in the way that the physical reality of its deployment would be. Kac asserts that, as an idea, the work could only have stimulated imagination, but by "actually, physically, materially" deploying the work, as intended, in space, "we are changing the very limits of art."

Ágora is a small holographic plate that, when illuminated by laser, displays the word "AGORA" (see Figure 4.16). The word "agora" translates as "now" in Portuguese, while the title of the work, *Ágora*, with its diacritical mark, refers to a public space. When the title and the work are received together, "space and time [are] bound in this two-word poem, articulated, like a hinge, by the accent mark." The hologram, encased in a titanium capsule (see Figure 4.17), was launched into deep space onboard the Centaur V, where it will adopt a heliocentric orbit. It will become the first artwork moving in perpetuity through deep space, the first artwork in orbit around

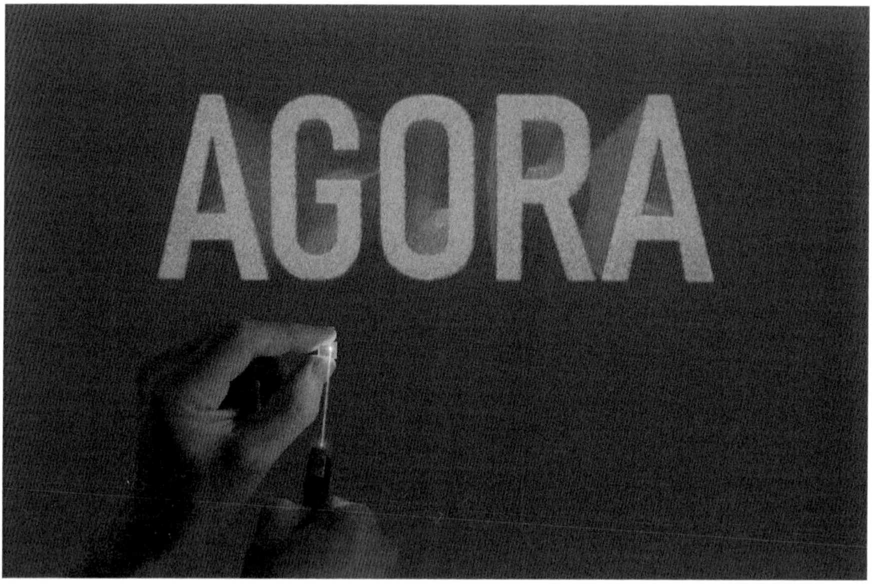

FIGURE 4.16 Eduardo Kac, *Ágora*, space artwork, 5/16 × 5/16" [0.8 × 0.8 cm], hologram in a 1.5 cm diameter titanium capsule, 1986–2023, moving through interplanetary space in perpetuity aboard the Centaur V spacecraft. The trimmed holographic plate is shown here illuminated with a green laser. Edition of 3 + 1 flown to deep space. Image courtesy Eduardo Kac.

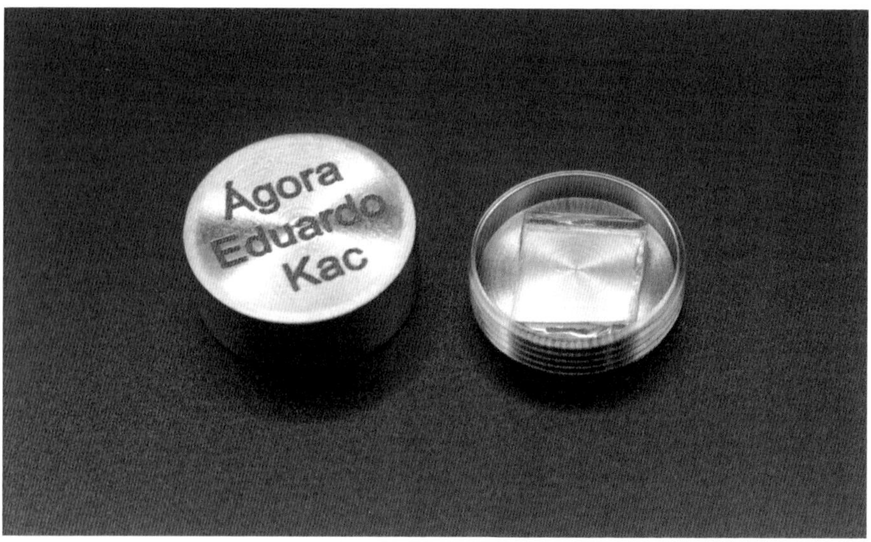

FIGURE 4.17 Eduardo Kac, *Ágora*, viewed in its titanium capsule, prior to the final integration into the Centaur V rocket. Edition of 3 + 1 flown to deep space. The title of the work is inscribed on the capsule lid in order that the poem and its title are not separated, hence, the two words "agora" and "ágora" remain paired. Image courtesy Eduardo Kac.

the sun, and the first transplanetary artwork, as its orbit takes it beyond the Earth and Venus. It is expected that the work will remain in orbit for "hundreds of thousands of years." Kac is optimistic that, in this time, the work will be encountered by future audiences who discover the Centaur V and investigate its component parts, much like archaeologists who have unearthed the treasures of Ancient Egypt.

Ágora is an edition of three, plus one "artist's payload" (a term Kac prefers to "artist's proof"). The three Earth-based editions have yet to be displayed at the time of writing, as Kac is still developing a system that will display them alongside a laser. No such laser has been included with the "artist's payload," so its message will only become apparent when future audiences expose it to the right conditions. Under the right conditions, the work is "a potential star." Unlike images that exist on a surface, or projections that are focused onto a surface at a set distance, holograms "continue to travel through space and time ... The message is spread out" across the universe, like light from a star. Kac notes that lasers are "integral to space exploration, occur naturally in the universe, and will certainly be available in the future, given their importance in so many fields, from engineering to medicine."

Kac has a forward-looking view of the destinies of both *Ágora* and *Télescope Intérieur*. He is unfazed by the knowledge that the work will initially be viewed by small audiences, intending for these works to have a lifespan far beyond his own. *Télescope Intérieur* "travels as a protocol," and can therefore be produced many times over. As tourist spaceflights become more common, the opportunities for this to happen will increase. *Ágora* will be seen by no one until manned spaceflights reach substantially further than they do at present, but Kac's intention is for some unknown future space traveler to serendipitously encounter the Centaur V, unhouse the work and deduce its purpose. Kac's attitude presents his works as founding artefacts in what may eventually be a long history of space art, that will eventually become an integral part of interplanetary culture. Throughout history, cultural artefacts have been preserved, or lost and rediscovered, and "this notion of cultural transmission is embedded in the fabric of society." Artworks are "transmitted through time" by the efforts of custodians and archaeologists, so it is highly likely, hopes Kac, that his work will have future spacefaring audiences.

Kac locates his work in the context of a distant future. "Even if it takes a hundred years," he says, "it is a fact that one day there will be a permanent base on the Moon," and interplanetary travel will become "ordinary." "We will expand our understanding of the human envelope of activity" into cislunar space and beyond. He draws comparisons between the state of spaceflight today with the advent of digital technologies as he experienced it in the early 1980s:

> When I started to work with digital in '82, I understood that this was the beginning of a new phase of human culture, and I felt that art and poetry

> had a place in this new culture ... It was more than just a new tool; it was a complete paradigm shift ... I feel that there is a similar process with space. We will see a similar shift. It will take longer, but it *will* happen.

We operate, now, at a stage in the development of spaceflight when we still view the phenomena encountered during spaceflight as novel, because "we still measure things in the universe relative to our frame of reference." As the spaceflight sector expands, and spaceflight is normalized, Kac argues that we will need to move beyond conceiving of all things relative to our experience on Earth. The first step in achieving this goal is to "exercise our imagination in these other realms" by engaging with, for example, weightlessness. For artists, such engagements should be how "we begin to develop a space culture in space, for space."

Eduardo Kac was interviewed on October 23, 2023.

5

Touching the Void

Performing weightlessness

Catherine James (2018: 3–4) suggests that artists' engagement with gravity conditions begins with their body. Exploring the ways that artists have employed gravity in their work on Earth, James observes that artists "intuit gravity," beginning with their body as a frame of reference. As a "primary condition of being," gravity is "mediated … through the body, rather than filtered through an abstract scientific discourse." This is, James suggests, partly the result of "our complete conditioning by gravity." Our bodies have been shaped by gravity and that body is our interface with the world, in a relationship that is also shaped by the gravitational force that grounds our bodies on the Earth's surface. Floating in the void, the body is ungrounded. The weightless body feels like an independent, celestial object, rotating around its own center of gravity. Just as astronauts experience space in a non-oriented way, without agreed upward or downward direction, the astronaut's body may also present at any angle relative to its surroundings or to fellow space travelers. Bodies may be encountered from above, below, or any angle in between, exposing views and body parts that are not commonly encountered in everyday social interactions on Earth's surface.

A greater appreciation for the bodily experience of weightlessness is one contribution that artists can make to the spaceflight sector, through examples like Ohad Fishov's *Nothing In Space* (2022), which imposes on an astronaut an hour-long pause from the itemized "doing" that fills their time on the ISS. Fishov's work "is a reminder that life is made up not only of doing but also of being," and that by so strictly structuring an astronaut's time in space around acts of doing, they are deprived of their humanity (Noble 2022: 151). While the regimented lifestyle of astronauts that leaves them with "little room for spontaneity, creativity, diversion, or play" (Ackermann 2016: 163), artists consciously make time for reflection and experience, provoking consideration of the phenomenology of spaceflight. This and other projects build upon the growing recognition that bodily experience of weightlessness underpins many human engagements with the environment of space. It is this realization that has informed the focus of programs such as CNES' *Résidence ZERO-G* (see Chapter 2), which mainly invites proposals that concentrate on the experience of inhabiting a weightless body. Their 2023 call sought projects to engage on a parabolic flight with the theme of ecstasy. Evidently informed by the observations made by previous residents, the call compares the states of ecstasy depicted in the visual arts to the sensory detachment from the terrestrial world that can be experienced in a state of weightlessness (CNES 2023b: 2).

Those who have produced or deployed artwork in microgravity speak about this activity as if it were a performance, some explicitly so. Frank Pietronigro (2004) chooses to describe *Drift Painting* as a performance, in which the novelty of the act of production is the main purpose of the work (see Chapter 2). The titular "drift painting" describes the action of painting while drifting, as well as the paint strokes produced during this activity; "painting" here is a noun and a verb. Onboard a parabolic flight, the moment of weightlessness is fleeting, and the work does not continue to exist in this state beyond the artists' actions of production. The artist is present in the work for the time that the work is active. Therefore, any viewing of this work must unavoidably include the artist and his actions. In this context, the characterization of the work as a performance further differentiates the work, as it exists in microgravity, from any artefact that returns to Earth following the flight.

The novelty of performing familiar tasks in new ways, including unfamiliar gestures and maneuvers by the body carrying out these tasks, highlights the act of production or deployment. Even where the artist is not present, the act of deployment in space becomes a moment of fascination. Yasmine Meroz and Liat Segal describe the difficulties that astronaut Eytan Stibbe encountered while unboxing and constructing their *Impossible Object* (see Case Study 2). Video footage recorded the complex maneuvers and unfamiliar actions that Stibbe had to perform in order to unbox and deploy their work, while also maintaining control of the camera that recorded that deployment. Meroz and Segal reflect on how unexpectedly different this deployment was onboard the ISS, when compared to rehearsals that had been carried out on Earth prior to launch, as a result of Stibbe's need to stabilize his body and the work in relation to one another, and in relation to the camera, and the impairment of motor skills that can result from weightlessness.

The weightless body moves in new ways. Becoming weightless, "all the preconceived ideas and notions about the placement of the body in space—the orientation of the self and everyone else—are suddenly rendered moot" (Benson 1999). For those unaccustomed to weightlessness, "each movement … leads to a totally unexpected displacement" (Dubois (2003 [1999]). Even trained astronauts who are acutely aware of the effects of microgravity and relentlessly trained for the physical actions that they must carry out in space, find that they must perform these tasks in unfamiliar ways. Actions must be relearned or substituted with different gestures that have been designed for the weightless environment. In training, astronauts learn a process of "translation," by which they become accustomed to new "action possibilities," creating additional significance in those actions above and beyond their output (Gast and Moore 2011: 323). For performance artists who work in microgravity, says new media researcher Annick Bureaud (2003), this is a "modification of the artist's grammar" with long-lasting effects in the way that the body moves even after a return to terrestrial conditions. Dancers and choreographers have sought to understand the effects of gravity on the body's movement, and the new movement possibilities resulting from weightlessness, both to inform their own practice and in support of astronaut training initiatives. The most significant of these is choreographer Kitsou Dubois, the first choreographer to experience weightlessness onboard a parabolic flight in 1990, leading to projects, including *Gravity Zero* (1994) and *Trajectoire Fluide* (1999), in which she has experimented and performed herself and, later, produced microgravity choreography for other dancers.

The mere act of embodying one's own form is defamiliarized in microgravity. Familiar actions such as standing or walking are not possible. The upright posture, or "anatomical position," that is representative of a default bodily configuration on Earth, is almost impossible to achieve in microgravity without artificial restraints. The weightless body instead adopts a neutral body posture that is part-way between sitting and standing, with bent knees and a hunched spine, arms floating forward and the head angled downward. Without the soles of the feet grounded on the floor, and therefore unable to walk or run, the astronaut must find alternative means of propelling themselves through space. On Earth, in the established model of the body and its interactions with our Earthly environment, "the hands [are] instruments of rational intelligence and feet [are] integral to the mechanics of bipedal motion" (Ingold 2004: 315). Feet are our interface with the ground, "becoming little more than a pedestal for the rest of the body," while hands have evolved "grasping and manipulation" abilities that make them adaptable to complex tasks, including gestural communication (ibid., 317). In space, however, where the body is not forced toward the ground, the hands play the role of grounding and transporting the body. In footage of the ISS, astronauts can be seen moving through the cabin propelled by their hands, grasping at handrails as they travel. When at rest, astronauts can occasionally be seen hooking their feet under the handrails, stabilizing their bodies not through their soles but through the top of their feet, grasping upward with their toes.

Dancers and choreographers report how experience of weightlessness helps them to better understand the relationship between bodily movement and gravity, and between the body and the ground where there are gravity forces grounding the body. Reporting for JAXA (2016: 3–13), Setsuko Ishiguro describes "dancing on the ground [as] a constant fight against the dancer's weight" or, says dancer Jeanne Morel, a "duel" between competing forces of gravity and muscle, in which the body pushes and the Earth pulls (see Case Study 7). The intuition that the human body develops over a lifetime of experience on Earth is, therefore, for all movement to be forceful. Supporting new researchers onboard MIT-funded parabolic flights, Sana Sharma observes how their instinct, when experiencing weightlessness for the first time, is to "push themselves off the ground," causing them to be propelled upward with unanticipated force, "then immediately stick to the ceiling."

Unaccustomed to weightlessness, new space travelers must overcome the fight to maintain the body "in opposition to … gravity" that determines how humankind instinctively approaches posture and movement (Dubois 2003 [1999]). When weightless, "you are not maintaining yourself in opposition to a force (gravity), but rather you let yourself be taken by the environment." Dubois prefers the term "gesture" to "movement" in the context of weightlessness because so little effort is required to generate a movement in this environment. She describes how "spinning around oneself is not a matter of moving mountains of muscles" but, instead, simply focusing "upon the gesture of moving your little finger for the rotating movement to happen" (Bureaud and Dubois 2005). For Morel, this feels more like an "impulse" than a movement, requiring little more exertion than a thought or intention that barely escapes the mind before it sets the body spinning.

This idea that spaceflight allows the full-body enactment of gesture is not new. Philosopher Vilém Flusser (2013 [1979]: 29–30) suggests that the human dream of flight arises in part from a desire "to be able to use the whole body as if it were a hand." The body, he observes, has been condemned "to the plane [the surface of the Earth]." As "prisoners of bi-dimensionality," pinned to the ground by gravity, humankind has aspired to fly in order to be able to move in three-dimensional space; to move our bodies in all directions the way that we move our hands; to become, like a hand, capable of three-dimensional gesture performed by the whole body.

Without this relationship to the ground, movement originates from the body's own center of gravity. Bodily movement in microgravity begins, not from the feet, but from the body's core: new gestures involving contraction of the pelvis replace directional forward-and-backward gestures (Dubois 1994: 60); movements beginning from the abdomen—yaw, pitch, and roll—are more naturally performed than any movement originating from the legs and feet, meaning that twists and somersaults are principal forms of movement in microgravity (Ahmadi 2020: 5). Dancer and choreographer Setsuko Ishiguro (2004) took advantage of the possibilities afforded by weightlessness to perform the actions and postures of *Hiten* (*Flying Deities*). Working from historic murals as the source materials for her various gestures and poses, Ishiguro imitated "depictions of … deities dancing and floating in mid-air." Onboard parabolic flights, Ishiguro found that "movements in mid-air resemble the movements of fish in water," as the body twists and bends in fluid ways that are different to the angular and directional movements that are performed when grounded. JAXA (2016) astronaut Koichi Wakata later performed Ishiguro's choreography onboard the ISS, enacting "curvilinear posture, translational motion, and rotator motion around a center of gravity."

Moving in these ways might be considered liberating, and, indeed, some do describe weightlessness as liberation from gravity (Murnik 2016: 67; Bureaud 2003; Ashcroft 2000: 246). Edith Ackermann (2016: 157; 166) describes the microgravity environment as "inherently playful," observing that astronauts take pleasure in performing gymnastic maneuvers that would be impossible on Earth. Some performances in microgravity embody this liberated, playful attitude. Ansuman Biswas and Jem Finer filmed their time onboard an Art Catalyst-funded parabolic flight, during which they played at being mythical genies, twirling around the cabin on flying carpets. The resulting film, *Zero Genie* (2001), "is simultaneously magical and hilarious" (Biswas and Finer 2005: 56). Marcel-lí Antúnez Roca (2005: 77) displays similar levels of glee in photographs of his performance for *Dédalus Project* (2003), in which the performance artist floats weightlessly while wearing a "bodybot"—a "wearable robot that controls the body." Antúnez notes, however, that the controlled environment of a parabolic flight, including "the caution of the flight controllers in releasing the securing ropes," meant that "instead of becoming a metaphor for lightness, the experiment became a paradox of control." The short periods of weightlessness that are afforded by a parabolic flight are temporary, and any loss of control is likewise temporary. There is always, in these short periods of playful loss of control, the reassurance of a return to gravity.

In more long-term experiences of weightlessness, as in orbit, there is no such reassurance. Without a ground on which to land, or friction to slow the body down, "trajectory is continuous," and a movement, once started, might never be ended (Bureaud 2003). A body, once it has started to spin, will continue to spin and spin for eternity, unless there is some collision or intervention. Dancer and choreographer Kitsou Dubois (2003 [1999]) describes her experience onboard a parabolic flight of "movements … without end. You have the feeling that the movement began before its preparation and that it continues after it." There is a kind of helplessness to this situation. Logan Smith (2022) observes that "astronauts on the ISS are constantly drifting and touching things to maintain their position relative to something," and to regain some control. When we lose agency over our own motion, Catherine James (2018: 5) argues that the human body "becomes object-like." In orbit, the body is in freefall, "a chaotic experience where the will is overtaken by gravity." For NASA, this necessitates a system of handholds that are always within reach; for dancer Morel, the only appropriate response is to give in to the loss of control; to enter a trance-like state, allowing the body to spin, and spin, indefinitely.

Spaceflight has taught researchers much about how the body perceives motion. Logan Smith (2022) observes that "we grew up with an intuition that is stillness," in which the Earth's rotation is normalized to the point that the human body does not experience it as motion. The human body is capable of perceiving acceleration and deceleration, or a change in direction, as the body's vestibular system detects and communicates these changes, but does not perceive constant motion except in the form of air resistance. It is for this reason that we do not feel the rotation of the Earth, and why astronauts in orbit feel that they are floating rather than falling. The experience of spaceflight teaches us that all motion and stillness is relative. The ISS travels at a speed of approximately 28,000 kilometers per hour, and astronauts inside the cabin appear to be still when they are traveling at this same speed. The phenomenon is helpfully demonstrated by NASA astronaut Jeff Williams during a reboost, performed to counteract drag and thereby maintain the station's orbit. During the reboost, the ISS accelerates. In a video, Williams is shown placing a camera so that it floats weightlessly, apparently at rest in mid-air, in the middle of the cabin (NASA 2013). When the ISS begins to accelerate, the camera moves forward, as if it is the camera that is accelerating. When Williams himself releases his feet from restraints, he too begins to drift in the same direction as if it is his body, not the ISS, that is changing speed.

The experience of movement and stillness is, then, subjective, in microgravity as it is everywhere. For Dubois, this subjectivity is at the heart of the experience of weightlessness, where verticality is subjective and movement must originate from egocentric references. Artist Bradley Pitts, whose work is explored later in this chapter, reports that the phenomenology of weightlessness cannot be captured on film or from an external viewpoint. Pitts describes how video footage "shows my body tumbling around and bouncing off the walls because of turbulence. But for me, any time that I wasn't touching anything, I felt completely stationary, with the plane tumbling around me" (Myoda 2013). For Pietronigro (2004) these contrasting perspectives are a temporal matter. He recalls working as if "in slow motion," while reports from his fellow passengers described him working at a "fast, frenzied pace." The inability to capture the phenomenology of microgravity is Jeanne Morel's reason for eschewing video in favor of other kinds of data capture, including brainwave recording, that might more accurately record the feeling of weightlessness (see Case Study 7). Morel stresses that, for audiences to be able to relate more effectively to the experience of microgravity, records of performance in weightlessness must record invisible sensations that cannot be captured in visual images. To experience space, she suggests, it is not enough to see it; you must feel it.

The void: Infinite and intimate spaces

Even looking up at the night sky from Earth we are able to understand space as an apparently infinite void. The word *space* carries connotations of emptiness and openness. What sets space apart from the open planes of Earth is that it is three-dimensional, extending outwards in all directions. Though it is not yet known whether the universe is truly infinite, the observable universe has the appearance of an endless void, and humankind's navigation of near-Earth space has allowed us to come face-to-face with this emptiness. The void of space is experienced as "the purest imaginable manifestation of the sublime," "characterized by vastness" and "darkness," offering "sensations of agoraphobia … and vertigo in terrifying proportions" (Smith 2009: 209). For artists, space introduces the possibility of occupying space in completely unrestrained ways. Bureaud (2003) describes this void as "complete space," being "limitless," three-dimensional space that surrounds the performer's whole body, so that movement can be performed in any direction.

Floating in the empty void of space is conceptually and physically different from flying through the air. Dreams of flying are founded on the idea of upward trajectories, in opposition to falling. Floating, by contrast, seems without direction and without risk. While we know that orbital trajectory is in fact falling, it is not experienced as falling down. In space, it is possible to "fall down" as well as "fall up" or to "fall away" in any direction (Du Preez 2019: 88). Unlike on Earth, microgravity is an environment where falling does not result in catastrophic collision, so that "fear of falling is not an issue" (Wightman 2005: 65). Of all the many fears that the space traveler might rightly experience in the hostile environment of space, fear of falling is not one of them. Onboard a parabolic flight, where most weightless performances have taken place, there is a kind of contradiction in which the controlled fall of the aircraft protects the body from experiencing that fall as such. At the time of writing, the *Résidence ZERO-G* program's current open call invites participants to engage with the theme of falling, focusing in particular on this "contradiction between freefall of the Airbus ZERO-G and liberation of the body from the constraints of Earth's gravity" (CNES 2024b; CNES 2024c). The call highlights that falling, in spaceflight, is an experience more akin to floating: a kind of suspension, in which the body is held rather than dropped; the weightless body is cradled by the space around it.

Acclimatizing to microgravity after nine parabolic flights, writes Dubois (2003 [1999]), becomes a matter of seeing negative space. Rather than fixating on surrounding surfaces and objects, she reports educating herself to see the "spaces between" as flexible spaces that can "dilate, contract, change their shape," and in this way, the body can learn to operate within the void and form a new kind of relationship with space. While locomotion on

Earth may be a duel between the bodily exertion and Earth gravity, in space it is a different kind of dialogue between the body and its surroundings: not one of competing forces, but of relative direction and speed. When one understands that weightlessness is not stillness but, rather, perpetual motion, it is possible to understand how the space between objects can be expanded and compressed through acceleration and deceleration, or by changing direction so that the body may converge with other objects. Thus, one is not reaching for an object; rather, one sets oneself on a collision course with that object. In this way, acclimatizing to microgravity means developing an intuition that space moves around the body as much as the body moves through space, and that there is no tangible distinction between these two phenomena.

The body in microgravity is not just flexible; it is permeable. Artists describe losing a sense of the boundaries of their own bodies, as if they are, themselves, "dissolving" into the void. Kitsou Dubois (2003 [1999]) describes, onboard parabolic flights, a feeling of disembodiment in which she loses a sense of her body's existence, only sure of the parts that are within her line of sight. This experience is familiar to astronauts who can "lose sense … of their limbs" when their proprioception is debilitated by weightlessness (Du Preez 2022: 109). The sensation of loss of physical presence extends to the body overall. Annick Bureaud (2003) describes how, "in weightlessness, we lose the awareness of the external limits of our body. It's as though the skin … stopped playing the role of sensor between the 'interior' and the 'exterior,' between 'me' and 'what is not me'." Similarly, Pietronigro (2004) reports that, during periods of weightlessness, "it felt like the boundaries of my body disappeared." Both describe the sensation of expanding beyond their own skin, as if the body merges with the void that surrounds it. In this "porous relationship between the body and its surroundings," the body is experienced as both weightless and boundaryless (Dubois n.d.).

If the body extends into space, then we can also imagine space extending into the body. At the same time as feeling disembodied, the performer embodies space. "Even the notion of self-identity as a body with biomass (weight) and mind is dissipated in zero gravity," writes artist Mario Ramiro (2000: 102). Ramiro describes the sensation of weightlessness as a physical embodiment of Descartes' principle, "I think therefore I am," in which "self-identity is but an identity of thought," as the sensation of physical existence disappears. For artist Bradley Pitts this is a new kind of emptiness in which the body is present and yet absent. Pitts experienced weightlessness onboard a parabolic flight for his performance piece, *Singular Oscillations* (2008, see Figure 5.1). During this performance, Pitts was naked, and kept his eyes closed and his ears blocked, so that he would experience nothing of his surroundings beyond the sensation—or *loss* of sensation—of weightlessness (Myoda 2013). He describes an act of immersion in the void, in which he

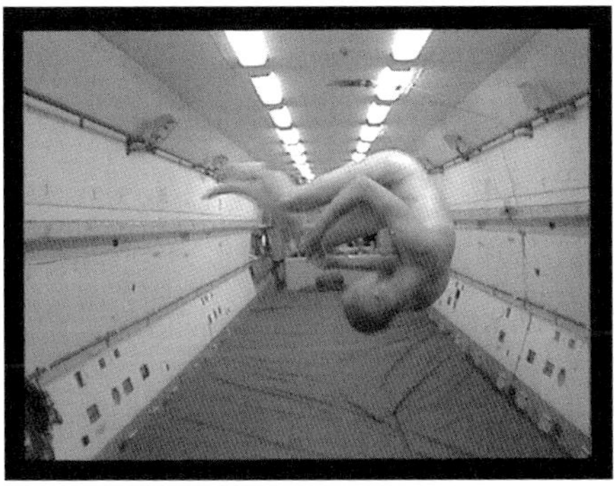

FIGURE 5.1 Video still from Bradley Pitts, *Singular Oscillations*, 2008. Pitts chose to be naked during his performance in order to enhance the loss of sensation that accompanies weightlessness.

passively allows his own body to be manipulated by space. Pitts' film, *Singular Oscillations: One Step Removed* (2008) is footage of his performance, from which the image of his body has been removed in post-production. The film shows the apparently empty cabin of the aircraft, its soft floor bowed by weightlessness, and flight crew interacting with some invisible object. The film represents the sensation of invisibility felt by Pitts during his flight, as if his body has become a part of the emptiness (Pitts 2011).

The experience of spaceflight is one of infinite and intimate voids. While space itself is portrayed as infinite emptiness (Gallagher et al. 2015: 5), the experience of weightlessness is at the body's surface. Astronaut Scott Kelly reports that, onboard the International Space Station, "skin comes into contact with other surfaces much less frequently than it would on Earth" (Harrington 2016). Clothes, for example, are a constant reminder of the presence of gravity. As Joanne Entwistle (2002) tells us, "dress has an intimate relationship with the body ... It lies on the boundary between self and other." This intimacy is caused by the "close proximity [of clothes] to the flesh," a proximity that is experienced visually and tactilely, and shaped by gravity. When weightless, clothes are lifted away from the skin, billowing outwards, surrounding the body but not draped upon it. Indeed, astronauts become so accustomed to living without the weight of clothes against their skin that some report physical discomfort from contact with everyday clothing after returning to Earth (Dunn 2016; Harrington 2016). For many, the first experience of weightlessness is lifting from their seats as they travel away from Earth. While they are still strapped in, the release of pressure from their backsides, and the tiny void created between their body and the seat beneath, provides haptic evidence of their weightlessness. In these ways, weightlessness is often felt in the minute voids that exist between the body and surrounding objects, rather than in infinite distance.

Mexican artist Nahum suggests that human contact is one way to provide relief from the void. Having experienced weightlessness as being "equal [to] nothingness" Nahum sought to restore a sense of presence through the intimate act of embracing fellow passengers. Floating through the cabin of a reduced gravity aircraft, Nahum recalls that, "It was when I saw someone floating in front of me that I extended my arms and embraced them. In the act of feeling the other, I felt myself. It was a reminder that we can only exist if we are together." In *Holding Air* (2015), Nahum performed a series of embraces onboard a Russian parabolic flight (see Figure 5.2). He describes how the experience of microgravity encourages a tighter and more intimate embrace than one would normally perform on Earth, as a result of reduced tangibility: "You notice that the body of the other person does not have any weight. It's like holding onto air. You get very anxious about that, and you try to hold on more" (Paterson 2015).

The above observations about the sensations of weightlessness have all been made within the climate-controlled environment of an aircraft or space vessel. There have not yet been opportunities for artists to perform in open space, where it is necessary to cocoon the body in a spacesuit. In this environment, the body contends with the void in different ways. Astronauts perform Extravehicular Activities, also known as spacewalks, wearing suits that house the same life-sustaining capabilities as a spacecraft,

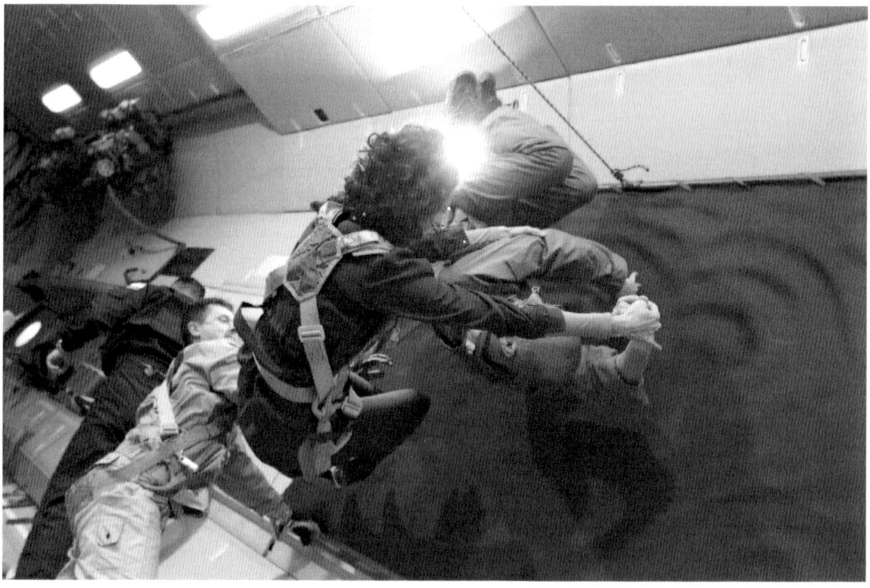

FIGURE 5.2 Nahum, *Holding Air*, embracing performance in zero gravity flight, 2015. Artist Nahum performs a series of intimate embraces while weightless onboard a parabolic flight. He likens the experience of embracing a weightless body to "holding air." Courtesy of Space Affairs. © Nahum

providing breathable air and regulating the astronaut's body temperature. The agoraphobia of being adrift within the void is felt in opposition to the claustrophobia of being confined within a space suit. The suit exerts pressure on the skin, and the biological functions of the body are viscerally apparent. The void of space is silent, and induces sensory deprivation, causing astronauts to become hyper-aware of the sounds generated by and inside their own suits (Berengueres et al. 2023: 190). They experience "sound and vibrations ... transmitted by the suit itself from the suit's contact with objects, or contact and friction between the human skin and the suit's inner shell"; meanwhile, recalls CSA astronaut Chris Hadfield, the sound of "your own breathing" is "your [constant] companion," layered with the sounds of the mechanisms that are working to keep their bodies alive—the fan and the liquid cooling ventilation that maintain air circulation and temperature (Berengueres et al. 2023: 191; Savage 2015; Frost 2012). In contrast to the vast visual landscape that the astronaut can see through their visor, this soundscape is characterized by its closeness, generated by and for the body.[1]

Counteracting the loss of pressure caused by weightlessness, the space suit exerts its own pressure on the skin, reminding the body of its boundaries. However, the thickness of the suit causes another dulling of the senses, making it difficult for astronauts to receive haptic feedback during their interactions with other objects. The suit creates its own void, a barrier between the wearer and any other object or surface, that, since it cannot be penetrated, feels infinite. Efforts are being made to reinstate the sensation of touch for wearers of spacesuits, including in MIT Media Lab's *SpaceSkin* project. This project is dedicated to the development of a space-resilient fabric that imitates the sensory capabilities of human skin, and includes *SpaceTouch* (2019–2022), gloves that map input from the exterior layer to a feedback system on the wearer's skin (SpaceSkin 2022). The technologies employed for this system are similar to those developed for haptic feedback in augmented reality and telehaptics (Yang et al. 2021). These developments create the potential for a spacesuit that enables the wearer to feel objects that are real or virtual; present or telepresent.[2] In these, the void is filled with tactile impressions of imaginary or distant objects that are not really there.

These experiences of contrasting aural and visual landscapes, and of objects seen but not felt, or felt but not seen, highlight the sensorial contradictions that characterize the phenomenology of spaceflight. The sights, sounds, and touch of space (or their absence) each tell a different

[1]Since no work has yet been carried out in this area, I offer readers a provocation: What is the sound art potential of a spacewalk? Where all sound is mediated by the spacesuit, could the suit become the site for a sound installation?

[2]Here again, there is scope for creative engagement but no existing practice for reference. I invite readers to consider the potential to use haptic technology in spacesuits to generate the sensation of an entire landscape of objects that are not really there.

story. What is seen is not what is felt or heard, and what is sensed by the performer is not what is seen by the audience. These contradictions arise in part due to sensory deprivation, in which the void plays tricks on bodies that are accustomed to the tactile sensations of gravity. Amanda Du Preez (2022: 101; 117) describes space as, above all, "unstimulating," writing that "the senses are highly mediated and lulled in deep space travel." For all that has been written about the physical and emotional impact of spaceflight, the actual sensation of spaceflight is overwhelmingly of nothingness: a nothingness so complete that extends into the body, so that the body itself becomes one with the void.

Making contact

On Earth, surfaces represent safety and stability but, for a body moving through space, they can become a hazard. Without gravity, "no longer can you trust anything" (Dubois 2003 [1999]). Weightlessness causes surrounding surfaces to propel the body away from them, as if rejecting human touch. Dubois describes unsettling interactions with surrounding surfaces during her first experience of microgravity, in which she found that with surfaces "if I knock them ... they push me back instantly. If I want to touch them, they also push me back ... With the floor, the ceiling, it is the same phenomenon." It seems that the natural order of things in space is to be alone, isolated from surroundings, lost in the void. Humankind has responded to a fear of the void by developing systems that reinstate a relationship between the body and its surroundings. Aiming to counteract the space traveler's loss of sensation and control, human interactions with weightlessness seek to manage chaos through systems that connect the body to surrounding surfaces.

Total detachment from ground or surface might be considered representative of human progress away from our Earthly origins. Tim Ingold (2004: 323) suggests that humans have evolved away from groundedness, referencing the chair, "a symbol of civilization"—and a move away from physical to intellectual inhabitation of space—that suspends the human body, ungrounding it. If the chair is symbolic of humankind's control over our relationship to Earth, then nowhere is human advancement more evident than in spaceflight, where the body is suspended in a void, grounded not through physical forces, but by artificial means, allowing us to control exactly how, and through what body part, we are connected to our surroundings. The impossibility of standing or sitting in space means that the space sector has needed to develop various artificial means of stabilizing astronauts' bodies. NASA's range of stabilizing mechanisms has ranged from seatbelts and tethers to foot restraints, thereby anchoring the body to its surroundings through a variety of different body parts (Thornton 1978: 19–26). In this

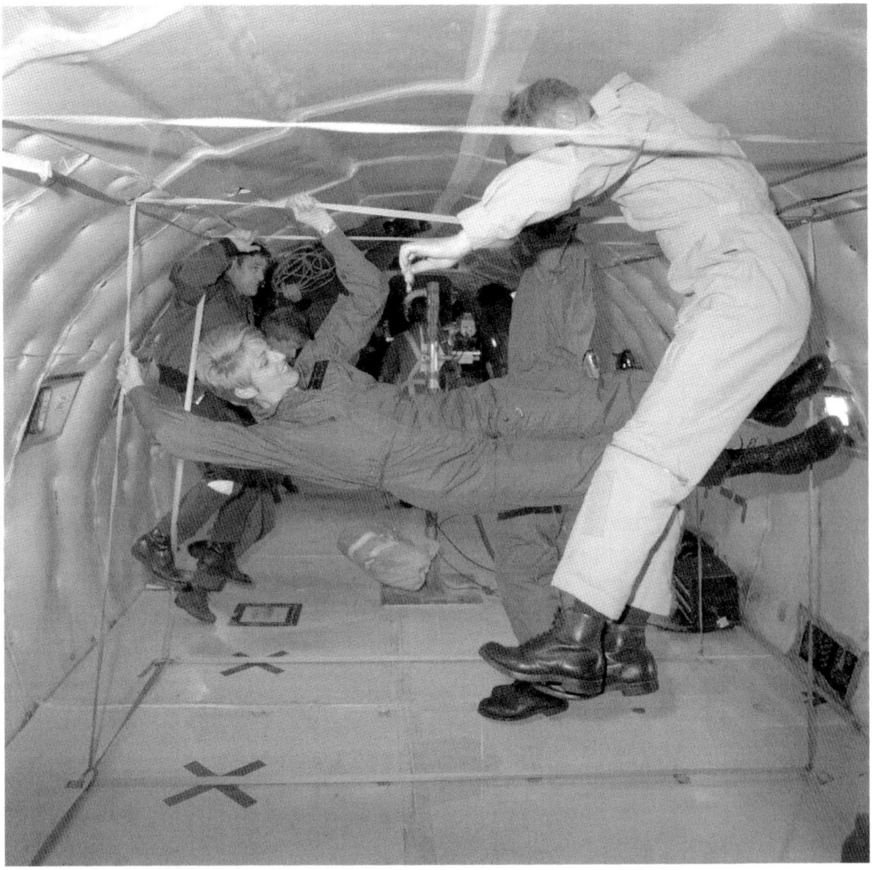

FIGURE 5.3 Vestibular Study testing in Zero-G flights on the KC-135, 1984. Seen in this view are Roberta Bonder and others experiencing weightlessness. Bonder (center left) stabilizes herself by grasping a system of straps, anchoring herself to the cabin with her hands while her feet float freely. Her fellow passenger (foreground, right) is out of contact with any tethers or surrounding surfaces, floating in the void. © NASA

way, microgravity reconfigures the relationship between the body and its immediate environment. The inescapable draw of the Earth on our bodies is replaced by a range of alternative methods of connecting the body to its surroundings.

As if in response to Ingold's ideas, Dubois has choreographed movement around a chair in microgravity. Onboard a CNES parabolic flight in 2004, Dubois and dancer Laura de Nercy performed Dubois' choreography around a black chair that was attached to the floor of the aircraft (Bureaud 2009: 5). Images of Laura de Nercy show her hovering above the chair as if she is hanging upside-down from the ceiling, with her orientation opposite to that of the chair. On the same flight, Dubois herself performed with the chair, anchoring herself to the back of the chair by holding it with her hand, then floating and twisting around it (see kitsoudubois.com/videos). Dubois has made efforts to replicate these reconfigured relationships between body and chair in performances on Earth, in extraordinary impressions of weightlessness including *La Chaise* (2018), in which de Nercy moves over and around the chair in a continuous movement that attempts to balance the gravity of the chair and the dancer's body. During the performance, de Nercy is grounded to the chair through her arm, shoulders, hands, and hips, but never sits upon it. The immense strength and balance that is required to perform these gestures on Earth is visibly evident in the tension in de Nercy's muscles. It is clear that she is using all of her strength to avoid slumping into the chair, as per the will of gravity. By contrast, in microgravity, these gestures are effortless.

Dubois, like astronauts onboard the ISS, avoids fully surrendering to the void by employing objects and devices. In many of Dubois' works, the body is grounded by grasping (see Figure 5.4). In *Perspectives* (2009), dancers perform on a parabolic flight alongside a pole and ring, both attached to the floor of the cabin. They propel themselves through and around these stabilizing devices. Morag Wightman takes a symbolic approach to stabilization in *Falling Without Fear* (2001). Wightman chose to attach herself by tethering herself with a rope and harness to the floor of the cabin of a reduced-gravity aircraft. The work "invert[s] the rules of suspension," so that, instead of hanging from a rope, the rope extends beneath her, as if Wightman's own body is bearing the weight of the aircraft (Wightman 2005: 62). The rope, and the act of tethering, is reminiscent of the way in which suited astronauts are tethered to the outside of their spacecraft during spacewalks, through an "umbilical." The umbilical serves the dual function of preventing the astronaut from floating away into space, while carrying life-sustaining power and air to their spacesuits (Du Preez 2019: 88).

These stabilizing, and potentially life-saving devices support concerted efforts to sustain a physical relationship between the body and other objects. Any kind of sustained contact with surrounding surfaces, writes Dubois (2003 [1999]), must involve a deliberate effort to "attach," to avoid being propelled back into the void. This necessitates making conscious decisions about how, and with what body part, the body is connected to its environment. Methods favored by the space sector involve grasping with hands and feet, and the ISS is lined with handrails for this purpose. This method of grounding requires both the action of grasping and the presence of something to grasp. Likewise, in Dubois' microgravity performances there is always a stabilizing device and choreographed gestures of attachment to that device. As Dubois tends to employ rigid devices such as a chair or pole, this means that the device becomes a pivot around which the body twists and turns. In other instances, as with Wightman's tether and harness, the device grasps the body. In the case of a non-rigid device like a tether, the body is

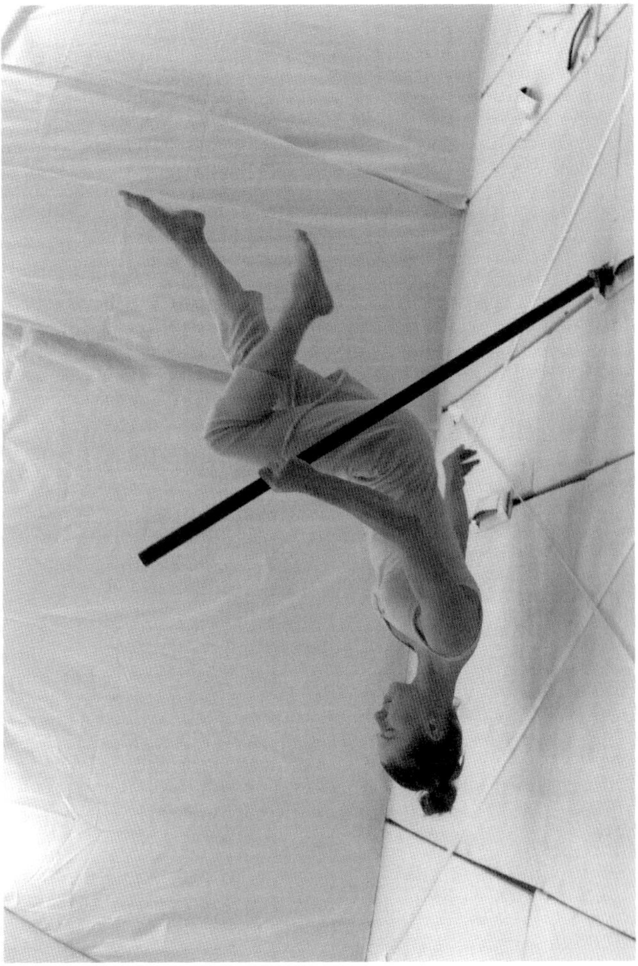

FIGURE 5.4 Kisou Dubois (choreographer), Chloé Moglia (performer), *Perspectives, le temps de voir*, dance performance onboard a parabolic flight, 2009, CNES Space Observatory. Photograph © Loïc Parent. A weightless performer stabilizes her body by grasping a pole, while no part of her body is in contact with any of the surrounding surfaces. Grasping the pole in this way the dancer is able to control her posture and relative position where otherwise, as other performers have recorded, weightlessness deprives bodies of control.

free to move around its own axis until it has drifted to a distance at which the tether becomes taught, at which point she feels the tug of the rope on her harness (Wightman 2005: 62). There is potential for performers to develop new systems and devices to ground or attach the body to their surroundings, that interface with any part of their body. Wightman's work demonstrates the potential to learn from existing practice in aerial acrobatics, where there is already a range of mechanisms for suspending the body, including by the mouth and the hair.

Velentina Sumini and fellow researchers at MIT propose that the body could be extended in new ways to enable new kinds of grasping and tethering. In *SpaceHuman* (2018–21), a tail-like soft robotics prosthetic device is worn to enable the body to grasp at its surroundings while the hands are kept free for other tasks. Inspired by seahorses, which use their tails to grasp objects, the prosthetic attaches at the coccyx, transforming the wearer into a tailed creature (Sumini 2021). In a different act of biomimicry, Xin Liu and Andrea Lauer's *Orbit Weaver* (2017) takes its cue from the web-building ability of spiders. *Orbit Weaver* is a hand-held device that "shoots out ... string with a magnet on the tip" (Ekblaw 2021: 34). The magnet can attach to any metal surface, and the string can then be retracted to pull the weightless wearer toward that surface. Both of these projects approach tethering devices as augmentations of the human body. Becoming post-human, the augmented body can begin to tame the void, counter-acting the loss of control that accompanies weightlessness.

Philosopher Elie During (2019) describes the augmented body of the astronaut as, on the one hand, "a diminished and enfeebled body," dependent on technology to perform the everyday tasks that are so easy on Earth's surface, and, on the other hand, "angelic" and "heroic," capable of achieving incredible feats. Augmentations designed for weightlessness, at the same time as allowing the body to perform in new ways—specifically, to perform in ways that are controlled and intentional—also highlight the body's incompatibility with the environment of space. For During, this is a matter of body and mind. While the body has employed augmentations to adapt physically to space, the human mind, he argues, is still grounded. Throughout the expansion of the space sector, "the brain has stayed home with us," and is as yet unable to process the experience of spaceflight "according to its own earthly frameworks." As a result, the augmented body of the astronaut is still "embryonic." We have yet to learn of its potential, and all of the ways that it might be further augmented to adapt to space. During views all spaceflight as an allegory for the augmented body. By sending ships and probes out into space, the human species seeks to extend itself in the same way as augmentations extend the individual human body. In this allegory, the relatively small proportion of space that has yet been explored by humans is akin to the limited understanding that we have of the human body's potential to operate within the environment of space.

Speculating about the future of humankind's interactions with the environment of space, it is possible to imagine a future in which we no longer feel the need to create stabilizing devices or to tether ourselves to surfaces, because we will have developed methods of operation that embrace, rather than resist, weightlessness. The potential for this is already seen in the way that artists and the space sector take different approaches to developing objects for weightlessness. The space sector's tendency has been toward control. The ISS, for example, is equipped with numerous devices to control and stabilize astronauts' tools, ranging from Velcro to slidewires. By contrast, many artworks that have been deployed in space are intended to float freely in the void. The artists eschew stabilizing or controlling devices in order to embrace the effects of weightlessness, intending their works to revolve freely within the microgravity environment: Arthur Woods references the choreography of an unrestrained, weightless dancer in *Cosmic Dancer* (see Chapter 4); Eyal Gever's *Laugh Star* (see Case Study 5) sets human laughter adrift like a celestial object. There is potential for the human body to be set adrift in the same way, freed from tethers and stabilizing devices. For this to happen, it would be necessary to develop the skills and mechanisms to safely move beyond the need to ground ourselves to a surface, and instead construct an approach to microgravity living in which all activities are designed to originate from our own centers of gravity. In this hypothetical future, it may become possible for performers to more fully embrace the notion of the human body as an independent celestial object.

Freeing the body would mean that we can develop new physical relationships with artworks, in which works can be grasped, navigated around, or otherwise encountered in ways that are relative to the viewer's body. Virgin Galactic requires that most human-tended experiments are designed so that they are wearable, tethered to the body so that they will not float away. Sirisha Bandla (2023) shares the example of conducting an experiment onboard a Virgin Galactic suborbital flight in which she, on behalf of a scientist on the ground, conducted an experiment that involved releasing liquid in vials. Had it been conducted on Earth, this experiment would have been desk-based, but for the microgravity environment the vials were tethered to her body and housed in a bag that was strapped to her leg. In this approach, artefacts become extensions of the body; the body becomes the stabilizing object or surface, and artefacts center around it. Sana Sharma's *Fluid Expressions* project demonstrates how this approach might be taken in the production of artwork. The second phase of her project included the development of a paint chamber that is tethered to the body and operated like an accordion (see Case Study 9). Sharma describes how her body and the paint chamber become a system, with a combined center of mass that is a short way beyond her own body, in the space between herself and the chamber. "It's a bit like I have a moon" she says, "I'm the Earth … We're both tied to each other … and we are both spinning around this central locus" (*sic*). In this system, every action that Sharma took affected both the cylinder and her own body. She likens the experience to being part of a Newton's cradle, in which momentum is transferred from one object to another. The significance of Sharma's project will be further explored in the next chapter, extending beyond how bodies interact with weightlessness to how weightlessness necessitates new artists' tools and techniques.

Case Study 7
Jeanne Morel, *Art in Space*, 2016–present

Dancer Jeanne Morel has performed on CNES parabolic flights since 2016, generating data for her partner, Paul Marlier, who interprets the data in visual and temporal forms for display to audiences on Earth. Morel's performances are "not just dancing in microgravity"; they are about "letting go," "being free," and achieving a "lightness of being" that is emotional as well as physical. These feelings are communicated to audiences on Earth through data visualizations that aim to capture the essence of the experience of weightlessness.

For many years, Morel has performed in extreme locations, from mountaintops to deserts. In these environments, Morel seeks to experience the fragility of nature, and of herself. Despite the need to board a plane to experience different gravity conditions, she considers her performances in microgravity to be a continuation of her engagement with the natural environment. Dancing in microgravity, she suggests, is "like a relationship with the sky … and sky is a big part of nature." Morel's movements are not choreographed. She enters a trance-like state, believing that only through intuitive movement can she achieve "a total connection [with] the environment." In these extreme locations, Morel feels that "dancing is too important to be choreographed." She asserts that, in order to sense an environment you must allow that environment to guide the body's movements (see Figure 5.5). The body must move responsively, intuitively, totally aware of the ways that the environment and the body flow together. In microgravity, this requires the body to sacrifice control.

Trance-like movement is particularly fitting for microgravity, where there is very little possibility of control. Indeed, most choreography rehearsed on Earth would be impossible to enact while weightless. Choreography is ordinarily "a duel between [competing] forces," balancing weight and resistance and, as such, it requires the body to exert force on its surroundings. Earth-bound movement must be "violent" and vigorous to compete with surrounding forces. By contrast, in microgravity, movements are "more quiet," more fluid, than those that a dancer must perform on Earth's surface. Rather than pushing or exerting forces, microgravity movement "is more poetic … If you want to turn you make an impulse," and that impulse, through almost no effort, becomes movement.

Having experienced several different gravities onboard her flights, including lunar and Martian gravity, Morel is certain that weightlessness has a far greater significance than any of these alternative gravity conditions. Moon gravity, approximately one-sixth of Earth gravity, feels like being "in a cartoon." In these conditions the body can bounce, hop, and do spectacular, playful things. While this is entertaining, for Morel it is not comparable to the sensory immersion that is possible in microgravity. She describes total

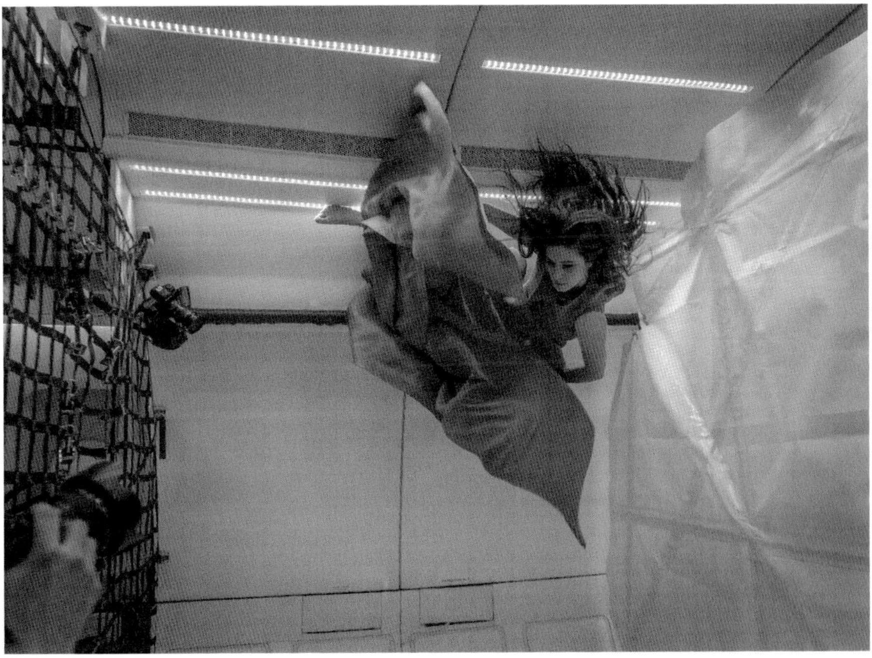

FIGURE 5.5 Jeanne Morel performs onboard a Novespace reduced gravity flight in 2018. Morel compares the sensation of weightless to flying, but stresses that "it's not like flying in a dream," rather, it's more a kind of "lightness"; "of letting go of something inside me." Image courtesy Jeanne Morel, Paul Marlier.

weightlessness as a "sacred space" that does not exist on a continuum with other gravity forces. Weightlessness is distinct, holistic, and profound in ways that reduced gravity conditions are not.

Her many experiences of different Earthly environments could not fully prepare Morel for the complete "lightness" of microgravity. Even underwater, where Morel has performed previously, "the ocean has weight so you can push"; "when you swim there is resistance; when you dance there is pressure." These external forces are absent in microgravity, where "it's just you and your body." Weightlessness is "more like a relationship with yourself" than with surrounding objects or surfaces; "all the impulses are with your own body." More specifically, these impulses begin at "my own gravity centre" (*sic*). She recalls how, when learning to dance, her instructors frequently referred to "*le centre du gravité.*" Drawing on this education, she allowed her movement to flow from "the abdomen and the hips." This is visible in video recordings of Morel onboard parabolic flights, twisting and tumbling around her core, stretching and contracting from her abdomen to extend and retract.

Morel's first flight, she recalls, was an extraordinary experience; a "revelation." During that first experience of microgravity she experienced "*une petite mort*"—a moment of extraordinary loss of the sensation of body and place, so that "I didn't know if I was alive" (*sic*). It was like "a rebirth ... Like I was in another time and space." She makes comparisons between Frank White's overview effect (see Chapter 1) and "the oversensing effect" that she experienced in her weightless body. It "was a new beginning. It changed my life on Earth, as a dancer, of course, but also as a human being" (*sic*). The "lightness of being" opened her mind to the possibility of "letting go of weight, letting go of sadness, [becoming] a better person." She was able to imagine herself discarding her physical self, to reflect on what she might be without physical presence. She asks what she might be without weight—"Am I Jeanne or not?," perhaps more "particles" than a solid whole; "Maybe I am just my feelings."

Morel's postgravity revelations include a newfound appreciation for Earth gravity. "Before," she says, "gravity was normal—we live with it every day," but "I didn't understand this relationship with Earth gravity before leaving it" (*sic*). She now describes a "love" for gravity that emerged only as a result of feeling its absence. While Morel continues to embrace short-term opportunities for weightlessness on a continuing programme of parabolic flights, she would not choose to experience long-term microgravity living. Making a hugging gesture, Morel describes gravity as a cosy, warm, familiar "friend" who will embrace the body and keep it safely anchored to the ground.

Subsequent flights could not live up to the profound emotional experience of her first flight. During an ongoing programme of parabolic flights, Morel has grown to be more strategic about her movements. Now more familiar with the conditions, she feels more able to engage in more calculated ways with the microgravity environment. Recent flights have become "more about working," with the goal being "to experiment with other ways to move [and] to create in microgravity." Morel and her partner, Marlier, have also had more time to consider the ways in which her experiences can be captured and interpreted. For the first flight, sensors were positioned on Morel's body "to draw the ephemeral dance of a dancer without weight" (see Figure 5.6). Their goal was to record the "flow" of the weightless body. Having been so affected by her first experience of microgravity, Morel concluded that it would be more fitting to record the mind as well as the body. It became important not just to record video of her dancing, but also to record the sensory and emotional feelings of weightlessness. On later flights the pair captured biometric data including pulse, blood pressure, cerebral activity in five fields, and the pressure on the body, brain, and heart. All of this data is visualized by Marlier in digital animations and sculptures. These abstract works reinterpret the data as undulating marks and textures (see Figures 5.7 and 5.8).

After the first flight, CNES enquired as to whether Morel would like to do an Earth-based performance with her body rigged in a harness so that she could replicate the movements that she was able to perform while

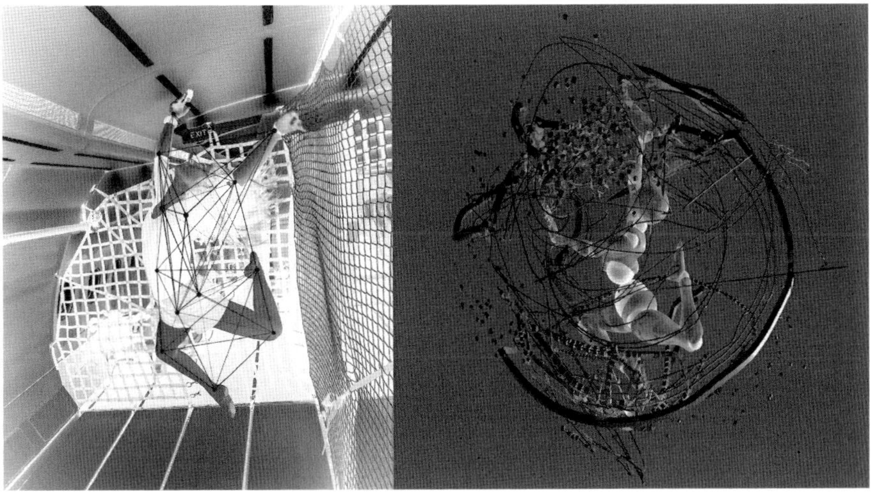

FIGURE 5.6 Paul Marlier and Jeanne Morel, *Danse en zerogravité*, 2016. Jeanne Morel performs in weightlessness (left) generating data to be interpreted as digital sculptures by Paul Marlier (right). Marlier's collection and interpretation of data are of personal significance to Morel as a result of her fear of flying. She describes needing to overcome, through training and meditation, a fear of flight in order to take up the opportunity to experience weightlessness. Her fear caused her to imagine what might happen if she were to die on her first flight, and she pictured a scenario in which her body would die but the data recorded on the flight would live on. Image courtesy Jeanne Morel, Paul Marlier.

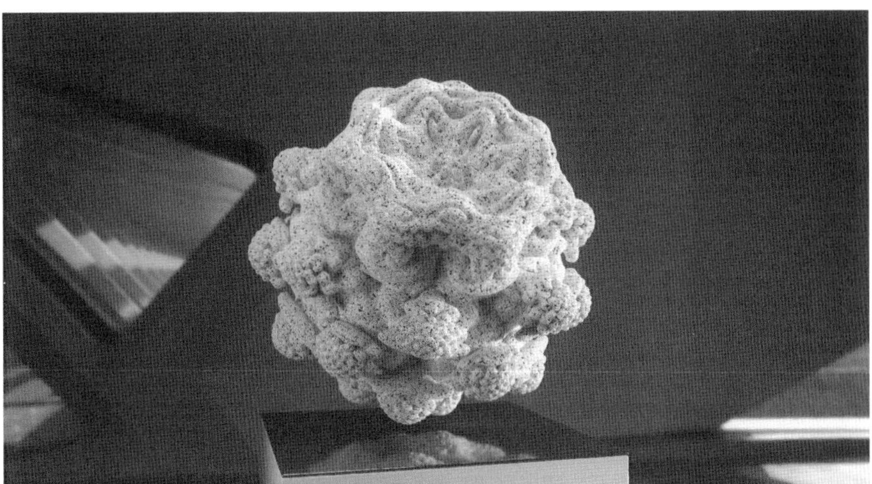

FIGURE 5.7 Paul Marlier and Jeanne Morel, *EOC*, brain data sculpture interpreting data from Jeanne Morel's performance in weightlessness, stone and resin, 93 × 93 cm, 2019. Marlier drew on his previous experience as a stonemason for this sculpture, aiming to "translate this absolute lightness, this total immateriality [into] something heavy, very heavy." Image courtesy Jeanne Morel, Paul Marlier.

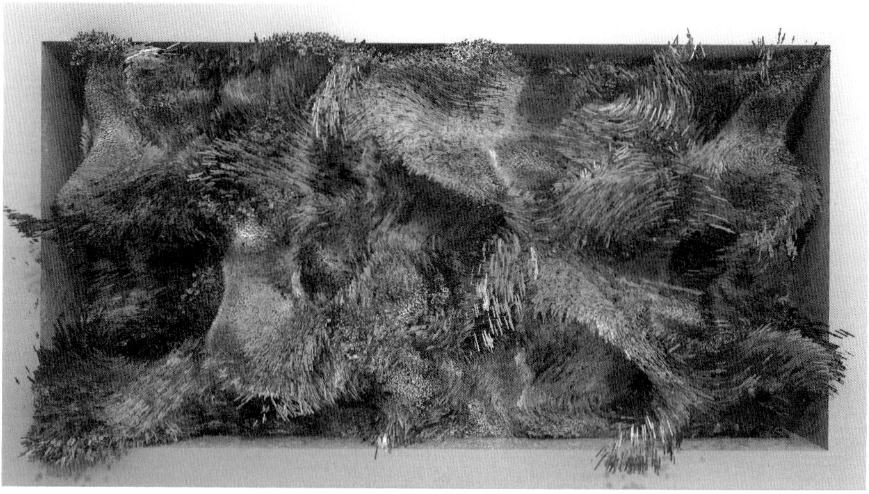

FIGURE 5.8 Paul Marlier and Jeanne Morel, *Renaissance*, generative data sculpture interpreting brain and heartbeat data recorded during a moment of trance during Jeanne Morel's performance in weightlessness, 2023. Image courtesy Jeanne Morel, Paul Marlier.

weightless. Morel rejected this suggestion on the grounds that it is important not to reduce the experience of weightlessness to aesthetics. While the body's movement in microgravity can be "beautiful," she says, the experience is "deeper," and it is these deeper feelings that she hopes to convey to her audiences. Moreover, to represent weightlessness through her own body, she feels, would be to exclude audiences who do not see themselves represented by her young, able-bodied, female form. It is for these reasons that many of Marlier's works are abstract. Data visualization "speak[s] about the invisible." It can depict sensations of "lightness" and "flow" in ways that are more abstract and universal. Marlier's flowing digital paintings are intended to be more immersive and profound than a direct depiction or recreation of Morel's movements, providing audiences with "an imprint of life" beyond Earth gravity.

Morel's ongoing collaborations with CNES have allowed her to develop a mutually beneficial relationship with astronauts, including her godfather, Jean-François Clervoy. She describes the sense of purpose that she experiences while collaborating with astronauts who appreciate the value of the different perspective that artists can embody in their work. In other sectors, including the press, however, she still encounters narrow views about the space sector. She stresses the importance of not allowing the military, governments, and commercial spaceflight companies to control the space exploration narrative. She asserts that "it is our responsibility as artists to speak about this quest ... to speak about what is important." What is important to Morel is that we do not view spaceflight as a conquest, but as an opportunity to learn to appreciate what we have on Earth. Referencing the overview effect, Morel stresses that engagements with space are not about focusing our attention away from Earth, but rather, they make us more conscious of the Earth and the ways that it is unique in the cosmos: "For me, Earth is the most beautiful dancer. She waltzes through the universe, and she is in balance, thanks to gravity."

Jeanne Morel was interviewed on November 23, 2023.

6

Making Space for Experiments

Space and craft

In order to establish a field of space art it is necessary to return to the crafts and tools that underpin material art practices. When art is intended for space, there are new factors—ranging from the effects of the harsh conditions of open space to the effects of weightlessness—that must inform an artist's choice of tools and media. Following a tradition of creative experimentation, artists' engagements with the environment of space have already prompted new approaches to the use of materials, and new methods and tools for the application of those materials. As outlined in Chapter 2, artists acquire experiential knowledge of how materials behave. Niedderer and Townsend (2014: 625) posit that experiential knowledge is intrinsic to craft practice, where "material understanding of making and haptic perception" underpin practice. Historically, this knowledge has been shaped by a lifetime of observation and experimentation in Earth gravity. Much of this knowledge needs to be set-aside when an artist seeks to create work for the environment of space, and new knowledge must be acquired before artists are able to predict the outcomes of their actions. Thus, experiments with materials in microgravity are an essential part of the artist's journey toward establishing the field of space art.

For artists who work in extreme conditions and inhospitable environments, outer space provides an opportunity to test the limits of their practice and push their materials to breaking-point. Japanese artist Azuma Makoto, who works primarily in "botanical sculpture," has launched floral arrangements and a bonsai into the stratosphere, suspended from space balloons. On Earth, Azuma is accustomed to displaying his work in climate-controlled environments, where temperature, light, and humidity can be optimized to promote the longevity of his work. However, his goal is not to preserve his plants indefinitely in their thriving state; rather, to extend the plants' whole journey from life to decay, recognizing that life is only significant if it is finite. He has shown interest in the ways that plants decay when exposed to different conditions and has explored hostile environments where plant life will not commonly survive, including saltwater, ice, fire, and the edge of space. *In Bloom* is a series in which flower arrangements are exposed to "naturally impossible situations," beginning with *Exobotanica I* (2014), later followed by *In Bloom #4*, *Exobotanica II* (2017). For *Exobotanica I*, Azuma launched payloads on two balloons, to a height of approximately 90,000 feet: the first, a brightly colored floral arrangement, and the second a bonsai tree (De La Cruz 2014).

Although they did not pass the Kármán line, and therefore cannot be said to have reached space, Azuma's works illustrate what happens to life when it is transported too far from the Earth's surface. Azuma categorizes his works as extra-terrestrial life-forms, "exbiota" (AMKK 2014), but exposure to the stratosphere soon deprives them of life. Extreme cold and wind make it impossible for the work to survive. The work is seen floating above a curved horizon against the black backdrop of open space in images captured by a Go-Pro camera that was suspended from the same balloons. The photographs make interesting companions to photographs referenced in Frank White's *Overview Effect*. The vivid colors of the flowers contrast with the darkness of space and the whiteness of the atmosphere below, reminding us that life does not belong in this environment. The fragility of life becomes even more apparent in images of *Exobotanica II* that show Azuma's arrangement breaking apart and flowers falling away from their stems. Christopher Hooton (2014) describes these fragments scattering into space, much like the debris from a failed rocket launch. The last moments of Azuma's plants' lives are significant because they are so fleeting. Art in orbit, similarly, often has a limited lifespan, burning up in the atmosphere as it de-orbits.

In the controlled environment of a spaceship or space station, life finds different ways to survive. "Biological properties change as gravitational force is diminished," meaning that microgravity has "profound consequences" for the growth and behavior of cells (Bizzarri et al. 2015). As well as exhibiting changes at a cellular level, plants reorient their growth (Baba et al. 2022). In this environment, the direction of growth of plants is determined by light more than gravity, meaning that there is potential to playfully experiment with it. These conditions enable works such as Kirsten Johannsen's *Nomadic Nature Kit* (2010), a microclimate sealed in two transparent domes that connect to create a spherical garden of plants growing in two different directions (see Chapter 4).

Inanimate materials can also appear to come to life, or behave differently, when weightless. Weightlessness can transform the properties and behaviors of materials in ways that affect their potential to be employed in the production of artworks, or the methods used in that production. Many common artists' media and materials adhere to surfaces differently when they are weightless, or form unexpected shapes. Wherever weight or gravity is exploited by artists on Earth, it cannot be exploited in the same way in space. Techniques including slumping, splattering, dropping, dripping, or hanging, are either not possible in microgravity or yield different outcomes. Similarly, the "peculiar behavior of fluids" in space is "completely different from our daily experience on the ground" (Monti 2001: 2). Wherever artists rely on fluid media, such as ink or paint, there is a need to transform expectations about how those media will behave when weightless or exposed to open space. In pursuit of an understanding of the kind of art practices that are possible in space, it is therefore necessary to consider the materials that are used in art production, or that may become usable as art materials when weightlessness changes their properties or behaviors. This, in turn, necessitates exploration of the craft practices that underpin the visual arts.

Experiments with these practices are important both practically and culturally as a stepping-stone in the journey toward opening up space as a site for arts practice. This book has already addressed how the democratization of space requires inclusion of all disciplines (see Chapter 1), including the arts. Ebru Kurbak (2023) argues that this is particularly the case for crafts, that are often "strongly associated with indigenous, nomadic, and women's cultures." For Kurbak, performing craft activities in space is one way to resist "the colonial instinct to import ... the latest technologies ... into new spaces." As was established in Chapter 2, assumptions are often made that engagements with space must be high-tech; and these assumptions are barriers to artists' participation in space where their work does not explicitly involve advanced technologies or involve collaborations with technologists. Craft provides ideal examples to demonstrate how artists' practice can be shaped by microgravity, not through the use of advanced technologies, but simply in the ways that craft can harness weightlessness through novel uses of materials.

ISS astronauts who choose to spend their leisure time engaged in arts and crafts have made discoveries that help to demonstrate how different established craft production methods and materials behave in space. Nicole Stott (2019) packed a solid watercolor kit so that she was able to maintain her hobby of painting during her time onboard the ISS. Stott describes the process of painting in microgravity:

> I would squirt out a tiny little ball of water from the drink bag and watch it float in front of me in zero gravity. Then I would put the brush toward it to touch it ... before I got the brush to the water, right before it made contact, the bubble of water seemed to move over to the brush, like it was attracted to the bristles in some way ... Then the ball of water would be floating around the end of the brush; it didn't mix with the bristles like it does here on Earth. I would then mush this water droplet around in the paint. But just like when I tried to "dip" the end of the brush into the ball of water, when I moved the ball of water toward the solid paint, it was like the paint was pulling the ball of water *toward it*. The

> same would then happen when I pulled the colored water back toward the brush. It was so weird to watch this whole dance of water and paint and brush take place. Now, finally, I would be ready to actually paint. But I had to be *really* careful as I was putting the brush to the paper, because if I got too close and actually *touched* the ball of water or the brush to the paper, it would just suck it up all at once. In the end, it was like I was just dragging this colored ball of water along the paper, just above the surface.

JAXA astronaut Koichi Wakata can be seen demonstrating calligraphy onboard the ISS in a New Year's broadcast (JAXA 2014). In contrast to the experiences reported by Stott, Wakata suggests that painting in microgravity is "convenient," because "even if you put too much ink in the brush, the ink does not drop." Kimiya Yui agrees, explaining that, when painting calligraphic strokes on Earth, gravity causes ink to roll downwards when the brush is overloaded, whereas onboard the ISS he is able to paint controlled strokes with his canvas at any orientation (Singapore Space 2017). Video footage of Yui's demonstrations show the astronaut loading coffee, used in place of ink, on to a brush before painting parallel lines on a piece of paper, and observing that these lines have the same appearance despite the orientation of the paper. Close inspection of the video shows the coffee loaded onto Yui's brush, pooling around the bristles, and then being pulled onto the paper by capillary action. The appearance of this transfer is very different to that which is familiar to artists on Earth, where the brush appears to suck ink or other wet media into its bristles. Yui's brush, by contrast, appears to act as an anchor to a globule of liquid that surrounds the bristles without penetrating them.

Outreach initiatives often make use of astronauts' personal art and craft hobbies in activities that humanize the ISS's mission, while also demonstrating the effects of microgravity through activities that are familiar to audiences. Karen Nyberg (NASA Johnson 2013) produced a star-themed quilt square as part of an outreach project onboard the ISS, that was later combined with other blocks submitted to NASA by members of the public from around the globe. Nyberg describes the challenges that she encountered attempting to judge and cut a square swatch with a quarter-inch seam allowance when preparing a quilt block onboard the ISS. The process of cutting and piercing fabric depends, on Earth, on the ability to lay fabric on a flat surface, and for the weight of that fabric to assist the process of cutting with scissors. These simple gestures are "tricky" to perform when fabric will not rest or drape. Nyberg found it necessary to adapt her methods, including adopting a cardboard template and taping her fabric to a surface.

Astronauts who dedicate time to arts and crafts both help and hinder the case for art in space. While they do, through no fault of the astronauts themselves, contribute to the problematic portrayal of art as merely a hobby or pastime, they also, to the benefit of artists, provide concrete examples of how microgravity affects art and craft production. These astronauts' experiences demonstrate not only that spacefarers have a desire to continue the craft practices that are familiar pastimes on Earth, but also that established craft production processes operate differently in space. Where astronauts such as Nyberg and Stott have found these practices difficult, there is evidence of a need to reskill, or to adapt tools and methods.

Contrasting these astronauts' activities with those of artists who have deployed work in space, the value of sending professional artists into space becomes apparent. Implicit in these astronauts' activities is an assumption that art and craft practice will adapt as minimally as possible to the environment of space, maintaining, as far as possible, familiar tools, materials, and processes. Although they do acknowledge, and are rightfully fascinated by, the effects of weightlessness on their craft practice, they typically view those effects as challenges rather than opportunities. Their response is therefore to make efforts to reinstate control or to focus on the degree to which they are able to reproduce familiar processes and outcomes. Their experiments therefore yield outcomes very similar to those that are already possible on Earth and, as a result, do little to contribute to progress in the field. When professional artists explore the potential for space as a site for art practice they tend to take a different approach, open to the possibility that space may enable new kinds of practice. Artists referenced throughout this book view the environment of space as an opportunity to reinvent their practice, to discard familiar processes, to explore new materials and, broadly, to create in ways that are not possible on Earth. They pursue what might be considered a post-disciplinary practice that either reinvents or imagines new methods of production.

"The magic of things going where no mathematics model has gone before"

By transporting familiar tools, materials, and practices to space, government and commercial space research initiatives have helped to discover and illustrate the extent to which space is different to the terrestrial environment. The ISS is dedicated to the study of microgravity phenomena, being principally a laboratory built for the purpose of enabling microgravity research (Fortezza 2001: 560). Microgravity experiments are conducted in order that they may "yield fundamentally new insight or discovery" that may benefit life on Earth or our future in space, and even now, after decades of research in orbit, "much is still unknown or not fully understood" (Rogers et al. 2009: 35; Chao et al. 2020: 7). Research in microgravity has recorded the strange behavior of familiar substances, and hence made visible some of the most significant differences between the environments of space and Earth. As a result of the novelty of their findings, microgravity experiments with familiar materials play a role in maintaining public and scientific interest in space exploration. New behaviors and properties observed in weightless materials reveal how little we know about the universe beyond Earth, including how little we know about the potential uses of everyday terrestrial materials when transformed by extra-terrestrial conditions. Astronauts are awed by "the magic of things going where no mathematics or space architect's model has gone before" (Battaglia 2012: 1093).

While experiments conducted onboard the ISS have revealed peculiar behaviors in the formation of cells, crystals, and flames, the most noteworthy experiments are those that involve liquids. Experiments exploring the behavior of liquids in microgravity inform almost all microgravity experimentation (Monti 2001: 1). This is largely because the behavior of liquids in microgravity differs so significantly in microgravity from the behaviors it exhibits on Earth, and because many other experiments also involve liquids in some way; for example, where there is a need to suspend cells in a liquid carrier (ibid., 2; 4). When isolated in space, liquid can become a "rigid body," achieving an equilibrium in which "its particles are subject to mutual attractions," pulled into a sphere by its own gravity (Jardetzky 2005: 4).

Liquid exhibits a number of specific behaviors in microgravity that contribute to it becoming sculptural. Small "capillary forces," more commonly known as "surface tension," become dominant (Martínez and Perales 2001: 21–2). While on Earth we associate *surface* with the *top* of a body of water, in microgravity, where there is no top or bottom (see Chapter 4); surface tension becomes a kind of container, surrounding a liquid on all sides, almost as if the meniscus becomes a perfectly fitted vessel. Those phenomena that we may encounter at a microscopic level when we interact with surface tension on Earth, become magnified, operating at larger scales. A single body of water takes the form of a "blob" or "drop," regardless of its size (see Figure 6.1). These may be "isolated" or "captive drops," either floating independently or attached, as a result of surface tension, to an object. If the drop comes into contact with more than one object, surface tension may cause it to become attached to both, with the liquid suspended in the space between those objects in a "liquid bridge" (see Figure 6.2). Martínez and Perales (2001: 22) also describe more complex combinations of these phenomena, in the form of "compound geometries" such as "drop inside a drop" or "a liquid bridge inside another liquid bridge."

Moreover, these sculptural forms have the capacity to undulate. In contrast to the ripples and waves that we are accustomed to seeing on Earth, that appear to spread only across the surface of a body of water, undulations in microgravity more visibly affect the shape of the entire body of water, combining "surface waves that [travel] around the outside of the sphere and body-centered waves—inertial waves—that [travel] through the middle" (NASA Johnson 2012b). Oscillations in shape and surface texture can be produced in numerous ways, including through the injection

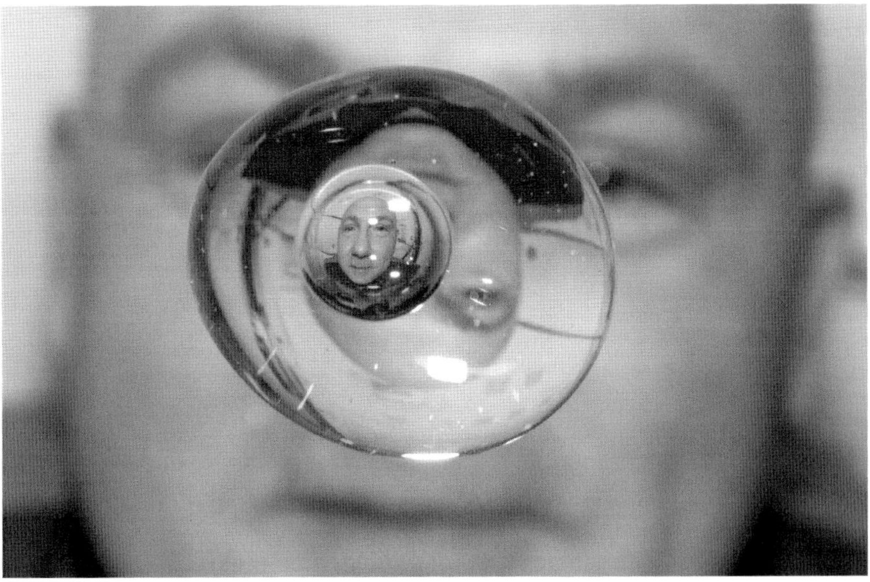

FIGURE 6.1 Astronaut Andre Kuipers watches a water bubble float on the ISS, showing his image refracted and reflected, 2012. One of several experiments carried out by astronaut Don Pettit, cited as an inspiration for artworks by Takao Fujiwara, Takura Osaka, and Sana Sharma. © NASA

of water or air, or by introducing vibrations through surface contact, such as by touching the globule with a vibrating wire. Soundwaves can also be employed to vibrate the surface of water globules, causing the whole form to jiggle and break apart. The phenomena vary depending on the size of the water globule and the tones to which it is subjected. Experiments by NASA astronaut Don Pettit, featuring hemispherical drops of water placed onto laptop speakers, demonstrate the extent to which the shape and texture of a water globule can change when it is oscillated by soundwaves: "depending on the frequency and amplitude, you can make anything from standing ridges to a ... series of lobes"; when the water globule is small, its behavior can become so "chaotic" that it begins to "squirt drops ... from all around the hemispherical surface" (NASA Johnson 2012a).

Artists including Takao Fujiwara, Takura Osaka, and Sana Sharma cite the experiments of Don Pettit as inspiration for their work. Pettit's experiments with water demonstrate the behavior of fluids in microgravity through visual records that are accessible to artists. Video-recorded experiments conducted by Pettit include demonstrations of how water clings to the surface of solid objects (NASA Johnson 2012b); oscillations introduced by puffs of air and water (ibid.) and soundwaves (ibid.); interactions between jets and films of water (NASA Johnson 2012c); how food coloring diffuses across water films (NASA Johnson 2012d); and how static charge on a knitting needle

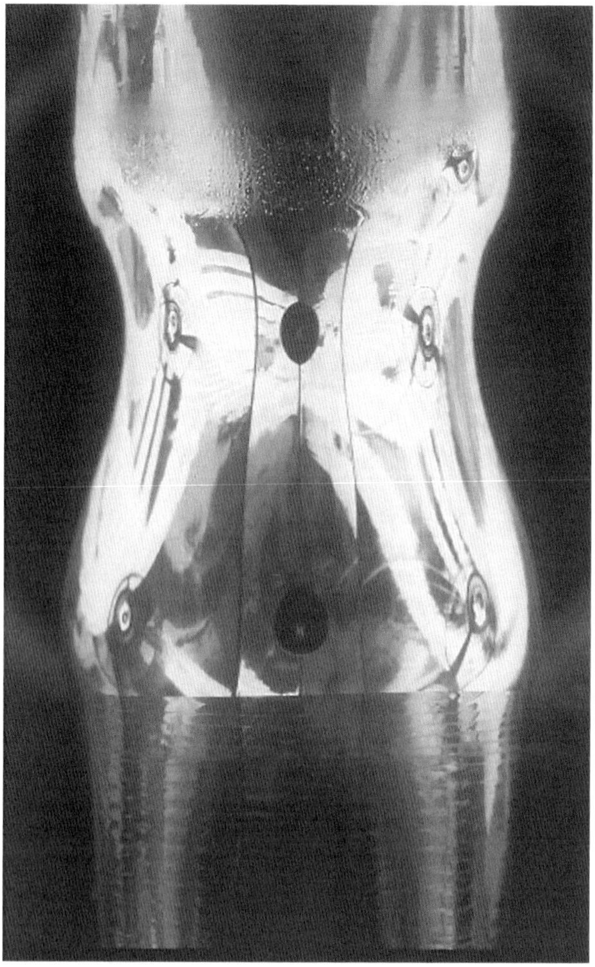

FIGURE 6.2 A liquid bridge is used to contain crystals during an experiment by Frank Szofran. The surface tension keeps the sample suspended between sample rods forming a liquid bridge, 1996. © NASA.

attracts water drops, causing them to orbit the needle (NASA Johnson 2012e). In the last of these, the droplets appear to have a life of their own, dancing around a needle, changing direction as if they are exercising agency, "like flies [at] a picnic" (ibid.). Videos and photographs of these experiments demonstrate to artists the many novel ways in which liquid media can behave when weightless.

These phenomena are significant for artists for several reasons: firstly, there are opportunities to employ liquid media that are not commonly understood as artists' media, including water, in new ways. Water globules much like those that form the basis for Pettit's experiments can be employed as a sculptural form, as in Meroz and Segal's *Impossible Object* (see Case Study 2). Pettit's experiments with static and sound reveal potential for kinetic artworks in which water-objects are moved or distorted by invisible forces. One such artwork is Takuro Osaka's (2003) *Sound Wave Sculpture*, in which a globule of "liquid and powder materials," trapped inside a cylinder onboard a parabolic flight, is "stirred by sound waves to change form and to produce a graphic pattern." Or, for two-dimensional artwork, a film of water might be used as a fluid canvas. The potential for this can be seen in Figure 6.3, an earlier experiment by Pettit that shows drops of food coloring suspended in a film, and in Figure 6.4, sugar crystals spread across a water film. Again, these works have the potential for kineticism. In Pettit's experiments with water films and vortices, food coloring is shown diffusing across a "two-dimensional sheet" of water, producing "mushroom cap" shapes and marbling effects (NASA Johnson 2012d). When a small puff of air is directed at the water, the marbled surface is set in motion. The colors dance and swirl and, since "the viscous forces in these thin sheets are small," the swirling continues for "five or ten minutes" without further intervention. Pettit's further experiments with "thin film physics" demonstrate how heat can be applied to the surrounding metal ring to create different kinds and directions of movement within the water film (NASA Johnson 2012c).

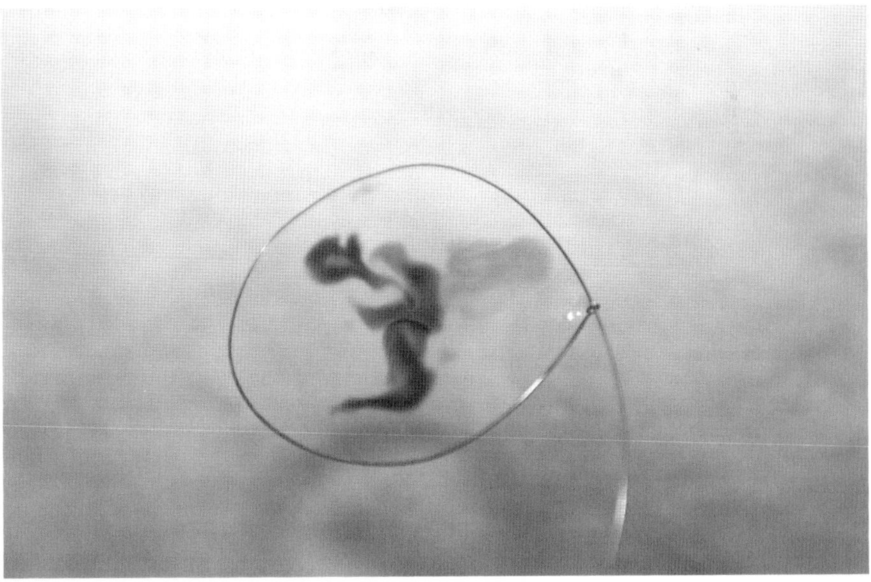

FIGURE 6.3 Demonstration of surface tension: a film of water is held in place by a metal loop onboard the ISS by Don Pettit, 2003. Food coloring has been added to the water. This experiment may inspire artists to consider the potential for a film of water to become an artist's canvas. © NASA

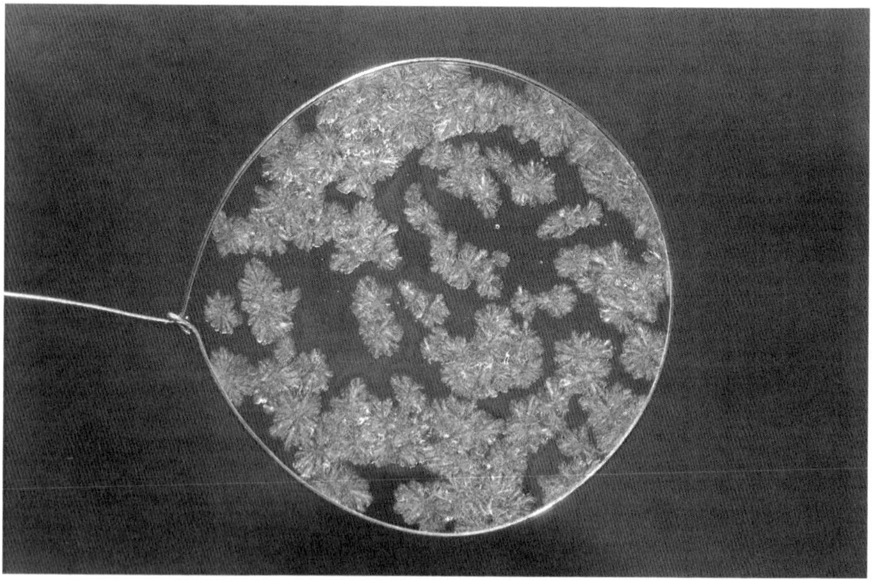

FIGURE 6.4 Sugar crystals on a film of water, suspended within a 50 mm metal loop onboard the ISS in 2003. Such experiments show artists the potential for a water film, acting as a canvas, to support solids as ornamentation. © NASA

Secondly, the peculiar behavior of fluids in microgravity means that established artists' media must be used in new ways. Artists' media are often liquid, and therefore behave differently in microgravity. There is a need to adapt any practices that involve liquid media, potentially involving the introduction of new tools and techniques. The behavior of floating paint in microgravity was the subject of research by the first British-born space tourist Richard Garriott. Onboard the ISS, Garriott released droplets of paint into a cube lined with white paper, observing that "as the paint dried it formed spheres rather than falling flat onto the paper" (Garriott 2008). Various artists before and after Garriott have played with floating paint in similar ways onboard parabolic flights, including Niu Bo in *Space Atelier* (1993), and Frank Pietronigro in *Drift Painting* (1998, see Chapter 2).

In recent years, artists have directly cited astronauts' science experiments as inspiration for their work. In *Artistic Experiments Using a Water Sphere* (2008), artist Takao Fujiwara generates deformations of a water sphere, using vibrations to explore "the possibility of new sculptural arts using surface tension" (JAXA 2016: 13). A sphere of water was held within an 8 cm wire loop, "two sticks made of wire with thread wound around ... were rhythmically inserted through the surface of the sphere to create vibrations." Images of the undulating sphere were captured for the purpose of disseminating the results of the experiment to would-be space artists of the future. Works by artist Takura Osaka, deployed on the ISS on his behalf by astronaut Gregory Chamitoff, employ the same metal rings seen in experiments by Don Pettit, used to anchor a sphere of water for *Marbling Painting on a Water Ball* (2007). In this work, ink and glitter are injected into spheres of water and stirred with a swab to create a marble-like effect. Two pieces of paper are then brought together on either side of the sphere, causing the liquid to transfer, producing two pieces of flat marbled paper. In a later work by Osaka, *Light, Life and the Sea* (2011), conducted on the ISS by JAXA astronaut Satoshi Futukara, the experiment is taken further. In a low-light environment, material produced by a bioluminescent species of "sea-firefly" and seawater, that Osaka (2011) had collected himself from Awase island, is injected into a 15 cm sphere of water. Paint is injected to add a cloudy effect to the water, and a black light is shone on the sphere so that the bioluminescent powder glows like a floating constellation of stars suspended in the water. The ink and fluorescent powder are seen swirling within the sphere, their movement originating from the site of the injection, making the sphere resemble the atmosphere of Saturn or Venus. This experiment is repeated with red, green, and blue, and combinations of the same, creating marble-like patterns in a series of water spheres. Osaka describes how the work prompted consideration of the interrelatedness of the micro and the macro, how the patterns in the sphere resemble Hubble-captures images of distant nebulas.

These works demonstrate that, when artists do not have the opportunity to explore the ways that their materials may behave in space, they can build directly upon the work of scientists. The research questions asked by artists working in microgravity are not dissimilar to those asked by scientists; Ansuman Biswas and Jem Finer's *Wave/Particle* (2001, see Chapter 3), for example, is underpinned by the question "what would liquids do in microgravity?" (Biswas and Finer 2005a: 54). The role of space art, at this time, is as much about learning what is possible as it is about making the case for a new field. Space art is often exploratory or experimental, and artists are often not able to fully predict the outcomes of their experiments. Indeed, much human activity in space, in all disciplines, is still exploratory and experimental. Artists, like scientists, are still exploring the ways that their materials are affected by weightlessness, and the new opportunities that the environment of space may offer to the arts. While so much is still unknown, craft experiments are necessary to build the foundations for the future emergence of a more fully formed field of space art in which approaches to microgravity art production may be more fully developed.

Experiments for future generations

Following Jan Woerner's descriptions of the four phases of space exploration (see Chapter 1), I posit that there are equivalent phases in the development of space art. In space art 1.0, art looked toward space, and artists produced astroart on Earth's surface; then, during the Space Race, in space art 2.0, artists proposed hypothetical projects for deployment in space, such as those proposed for the 1986 Eiffel Tower Competition. Now, in space art 3.0, artists are beginning to learn what might be possible through foundational observations and experiments. Without sufficient experience of how their materials will behave in microgravity, artists must necessarily acquire new knowledge and develop new tools and approaches before they are able to progress to space art 4.0, a fully developed, new field of art associated with an established set of practices. In space art 3.0, there are three main factors at play: the fact that the environment of space is still novel to humankind, and so all engagements with space must be, to some degree, experimental; that access to space is still dominated by other sectors; and that artists and scientists have different understandings of what constitutes research (see Chapter 2). The consequences of these factors are that space art in the current era is often focused on craft and materials, and necessarily experimental in nature, sometimes yielding unexpected outputs; that artists still have much to learn from experiments conducted by non-artists, including astronauts such as Don Pettit; and that artists are still struggling to assert their right to learn in the intuitive ways that are standard in their discipline. These experimental practices and observations are necessary in forming the foundations of a new field of space art.

Artists exploring microgravity often describe their work as foundational or experimental, expressing a collective understanding that the main purpose of their space art practice is to acquire new knowledge for the benefit of future generations of space artists. Takao Fujiwara stresses that his "artistic experiments" do not constitute a complete artwork; rather he emphasizes that "the outcome of this research will be a foundation for experiments in art using liquid in the future" (JAXA 2016: 3–10). Fujiwara's work is similar to, and directly informed by, experiments of astronauts including Don Pettit and JAXA astronauts Mamoru Mohri and Takao Doi, but with the stated goal of ascertaining potential methods for the production of future liquid sculptures, and acknowledging that there is still further work to be done before artists are able to understand and exploit the full potential of this method. Fujiwara is not alone in choosing to present his work as an experiment for the benefit of future generations of artists. A 2016 JAXA *Report on the Pilot Missions of Utilization for Culture/Humanities and Social Sciences* that summarizes the impact of works by Fujiwara and other artists refers to these as "pilot missions," presenting these projects less as artworks in their own right, rather as initial investigations into the potential of space as a site for art practice, the methods and materials that might be employed in future works, and the ways in which these might involve other stakeholders. The report's many conclusions include observations about how the necessary involvement of astronauts in the deployment of experiments forces collaboration across disciplines. The report also makes a case for continuing to invest in these collaborations, based on the interest shown by Earth-bound audiences in creative work that has been deployed in space.

In the examples highlighted in the JAXA report, there are close similarities between the experiments of astronauts and artists, with artists often employing tools previously used by astronauts. Takura Osaka (2017) acknowledges the similarity between his own experiments and those conducted independently by astronaut Gregory Chamitoff. He stresses, though, that artists add value to these experiments that astronauts cannot. He asserts that there are differences between similar water-sphere experiments conducted by astronauts and artists resulting from the artists' ability to visualize color combinations during their design of such experiments. Osaka suggests that his training in the arts has taught him to visualize and predict the ways that colors would mix together in a water sphere, and to therefore take a purposeful approach to color selection. Astronauts, he claims, without this training, will mix colors without knowing what combinations they will produce. Indeed, this is seen in videos of experiments conducted by Don Pettit, in which Pettit is seen continuing to mix colors until they turn black (NASA Johnson 2012d). This suggests that, through a combination of artists' training and insights gained by watching astronauts' experiments, it is possible to make progress toward a predictable outcome, and perhaps one that might constitute a completed artwork.

Osaka describes a relationship between astronauts' and artists' experiments that is reciprocal and iterative, with each contributing new knowledge that will be beneficial to both disciplines. Small steps are made toward developing a shared understanding of the behaviors of materials in microgravity, and the ways in which these behaviors can be exploited in any discipline. Iterative experimentation across disciplines, leading to incremental development of new knowledge, is one way in which multi-disciplinary teams can better understand the value of each other's work (Tolfo et al. 2024). This approach therefore has the advantage of helping to demonstrate the value of the arts to the space sector, as well as enabling artists to take advantage of knowledge generated outside their field. For artists, building directly on the work of science experiments may involve redeploying tools or techniques that have previously been employed in science experiments, or simply adapting a scientific experiment so that it produces different outcomes.

It is important not to undervalue the artist's knowledge-production activities in this iterative, cross-disciplinary knowledge exchange. For multi- and interdisciplinary artist Alana Jelinek (2020: 11), both the arts and sciences are knowledge-forming disciplines. Jelinek (ibid., 1) observes that, traditionally, "art has been understood and valued as a form of self-expression and … not as knowledge producing" and argues that we need to recognize that artists learn and further discipline-specific knowledge by self-consciously building on the achievements of other artists. The activities of artists in space suggest that we need to return to the idea of art as a knowledge-forming discipline. In the context of spaceflight, where science and engineering dominate, where the art/science relationship is assumed to be the main disciplinary context for arts practice (see Chapter 2), and where art experiments are often directly informed by scientific experiments, there is a need to assert that art can, independently of any scientific experiments that have informed it, create new knowledge. Moreover, it is important to hold onto the principle that new knowledge produced by artists in the environment of space does not always have to benefit other disciplines; rather, it can be mainly or solely for the benefit of the arts, and this makes it no less valuable.

Most examples of art in space could be described as "practice as research" (Nelson 2013). It is with the characterization as "research" that most projects to date have been funded. Arguably, this praxis is discipline-forming, contributing to defining the new discipline or field of space art, identifying its concerns, establishing the tools, techniques, and media that distinguish it from terrestrial art. In the first instance, this research and the new knowledge that it generates is for the benefit of other artists. It tests the potential for materials to be employed as artists' media, trials new methods of production, and broadly works toward an understanding of what art practice is possible in the environment of space, so that future artists may deploy those same media and methods more confidently. With this in mind, it is worth returning to the question of the intended audience for space art, as raised in Chapter 3. If we view space art, at this stage in its development, as so experimental that its main aim is to lay the foundations of future practice, then it follows that its main beneficiaries are other artists.

Space artists are following in the footsteps of artists who have pioneered new tools and media in terrestrial art, such as light art or digital paintings, which both enabled artists to paint with light instead of pigment (Sethi 2022). Comparisons may be made to Man Ray's pioneering light paintings, or Vera Molnár and Manfred Mohr's generative art. A significant difference between these historical examples of innovation that directly employ new technologies as media, and the advancements made in space art, is that space art exploits the technology of spaceflight to grant access to novel behaviors in existing media. Just as digital art replaced the brush with the stylus, microgravity conditions may make common artists tools such as paintbrushes obsolete as artists develop new tools and alternative methods of painting or sculpting. Existing tools are deployed in new ways, tools are appropriated from other disciplines, or entirely new tools are developed specifically for the weightless environment. For example, in *Drift Painting*, Pietronigro (2004) employed pastry bags as his main painting tool, and "attached magnetic loadstones to 'wands' and planned to use magnetism to help move the floating metal infused paint." For Pietronigro, this choice of tools helped to counteract the formality of working with NASA.

Where existing tools are lacking, further progress toward establishing the field of space art may be made when artists develop new tools that enable new methods of production. MIT designer and researcher Sana Sharma observes that, since paint behaves differently in microgravity, new tools need to be developed for its application. Inspired by Don Pettit and Nicole Stott, Sharma has sought to reinvent the way that paint is deposited onto canvas. Having observed that bristles do not absorb liquid media as they do on Earth, Sharma has chosen to do away with the brush and explore other methods of depositing paint onto canvases, including injection systems and metal rods. Sharma's project is, like the activities of Don Pettit, a series of experiments. Her primary goal was not to produce artwork, but rather to explore the ways that other future artists might themselves produce new kinds of art. The legacy of her project will therefore not be her canvases, but the ways in which she has challenged terrestrial methods. The observation that the paintbrush—the icon of terrestrial painting practice—may become obsolete in a spacefaring society, illustrates the extent to which spaceflight has the potential to transform existing art practices.

In their experiments, artist-researchers such as Sharma, Osaka, and Pietronigro are determining the nature of future space art. Where they develop new tools and techniques, they are potentially also determining its aesthetic. Systems such as Sharma's *Fluid Expressions* (see Case Study 9) and methods demonstrated by Frank Pietronigro, will inevitably lead to a style of space painting that is dominated by splatter-style marks and chaotic compositions that are more the result of the environment than of the artist's intention. It will be up to future artists to escape this technological determinism, and, if so inclined, to demonstrate the potential for these tools to be used in the production of more idiosyncratic work, with more deliberative output. These artists' output has much in common with terrestrial experiments with materials that have, in their own time, been considered "radical"—processes that involved dropping, spilling, and splattering, such as the works of Jackson Pollock or Lynda Benglis (Frascina 2000: 120; Richmond 2015). Aesthetically, space artists do not take such a radical step away from existing art practices as these other artists, but the unfamiliar behaviors of their materials in the environment of space mean that their processes are novel.

Inevitably, while weightlessness is still considered a novelty, its effects will be fetishized. When any new technology is first adopted by artists, early-adopters' "work is ultimately a showcase of the technology itself" (Colman 2005: 279), and the same may be true of recent spaceworks. These projects must explicitly describe the process of production, so that audiences are able to appreciate their significance, and so that future space artists may learn from them. As materials behave so differently in microgravity, space art 3.0 inevitably foregrounds this behavior. It remains to be seen how future artists will advance the field in space art 4.0, but in order to empower that future generation of space artists it is necessary for the current generation to demonstrate what is possible.

Case Study 8
Alexis Hope, *Space/Craft*, 2019

Aboard a parabolic flight, Alexis Hope expelled hot, colored glue from a glue gun to produce fluid forms in microgravity (see Figure 6.5). Hope floated outside a transparent containment cube that housed an arrangement of small balls on sticks, arranged in imitation of a constellation, that could provide anchor points for trails of liquid glue. The containment cube was fitted with openings, through which the artist was able to reach to create her work. During the flight, Hope and her work were exposed to a variety of gravity conditions, including microgravity and hypergravity, each of which affected her experiences of performing *Space/Craft*. Hope used a hot glue gun to shape forms during periods of weightlessness. At the end of each short period of microgravity, Hope's sculptures, not yet cooled and solidified, sank and splattered on the ground.

Hope's medium of choice, glue, is encountered on Earth in both solid and liquid states. Its potential to transform from one state to another is essential to its function, and the transition between liquid and solid familiar to regular users of glue guns shapes artists' interactions with the material. Hope chose glue as her medium aware that, in space, liquid can resemble a solid. *Space/Craft* responds to a transformed understanding of how liquids behave that is achieved as a result of observing this behavior in microgravity. Hope approached the project curious about the "ephemeral sculptures" that might be possible when hot glue is released in a microgravity environment, where the glue could have the properties of a liquid and a solid, being simultaneously fluid and sculptural.

For Hope, the decision to use glue as her medium was partly driven by its inclusivity. As a community organizer who conducts participatory design practice, in environments were there are "low barriers to making things," Hope strives to use "accessible materials" that are often "low tech" and "messy." While *Space/Craft* invites comparison to Pietronigro's *Drift Painting* in its execution, Hope's use of craft materials ensures that her project is distinguished in ways that are important in the broader context of commercial spaceflight. Despite efforts to democratize space, the commercial spaceflight industry is still perceived as predominantly male, and, due to the high cost of participation, space is an exclusive destination that is inaccessible to tourists without significant income. Craft materials, with their connotations of accessibility and femininity seem out of place in this context, and highlight preconceptions about the profile of individuals who engage in spaceflight.

Emphasizing the accessibility of her tools, Hope stresses the importance of her choice of a commercially available glue gun. In her interactions with others in the MIT Space Exploration Initiative, she encountered the view that a specially engineered "space-glue gun" might be more appropriate. Hope concedes that, indeed, if the project had been more long-term, involving

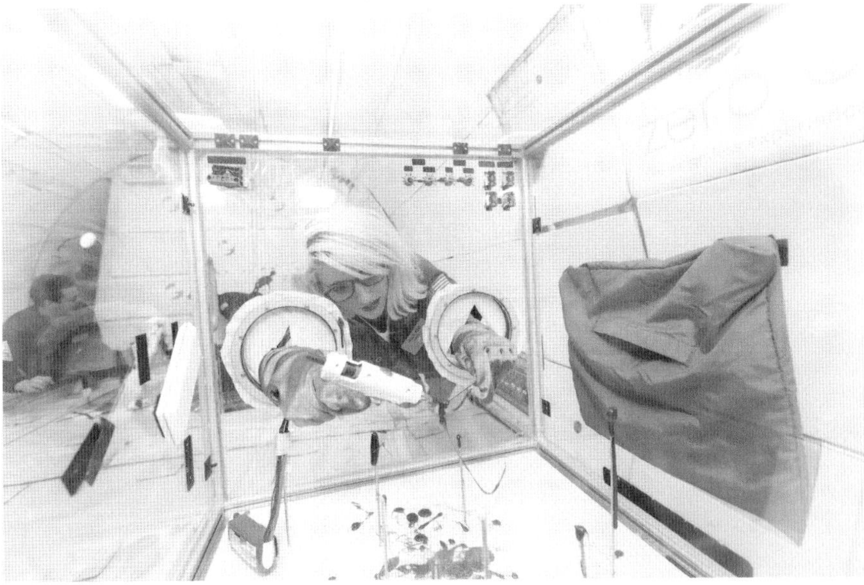

FIGURE 6.5 Alexis Hope, *Space/Craft*, colored glue, glue gun, wooden sticks, polystyrene balls, 2019. Hope performs *Space/Craft* aboard a reduced gravity flight. The glue gun is sealed within a containment cube, operated by the artist from outside the cube as she floats freely, having detached her legs from the straps intended to secure her body to the cabin floor. Ephemeral sculptures are formed by the hot glue in microgravity, anchored to an arrangement of sticks and balls that represent constellations. Splatters can be seen on the bottom of the cube, where the glue has fallen to the floor in the transition between gravity conditions. *Space/Craft* was enabled by MIT's Space Exploration Initiative and took place onboard a reduced gravity flight operated by ZERO-G. Image courtesy Alexis Hope, © Steve Boxhall / Zero-G.

multiple flights, she might have sought to develop a glue that would solidify more quickly and thereby retain the shape that it adopts while weightless. For her first and only flight, however, Hope's use of off-the-shelf tools and materials positions her work alongside other contemporary art practice that shares the goal of elevating craft. With materials that were not intended for microgravity, Hope was able to embrace the unexpected behaviors of those materials in this unfamiliar environment. This loss of control, over her own body as well as her materials, is of particular interest to Hope and a core part of her intention for the project.

When framing her project for MIT, Hope stressed that the work would focus on experience more than any material outcome. Unlike Pietronigro (see Chapter 2), Hope did not have to re-frame her project as a science experiment in order to receive support to execute it onboard a reduced-gravity flight. Hope describes the MIT Space Exploration Initiative as being "interested in art for art's sake," and as such the institution was willing to support a project that was explicitly presented as craft. Nonetheless, there was a requirement to describe the project in "utilitarian" ways, and to identify a physical outcome. Prior to the flight, Hope carried out trials to test what it might be possible to sculpt during the 15–20 seconds of microgravity that she expected to experience during each parabola. At the moment of execution, she found that this was not so straightforward. The lack of control that she experienced during microgravity, combined with the need to adjust to changing conditions at the start and end of each parabola, and the unexpected behaviors of the weightless glue, meant that her time was dedicated less to the act of production than to the act of learning. The glue did not behave as it did in her experiments on Earth. It would "ball up and bunch up" in unexpected ways. In each period of weightlessness, Hope learned more about the effects of weightlessness on her material and how it could be shaped. While this might otherwise have been presented as an investigation into the properties of the material in microgravity, Hope recognized that this would not have been sufficiently novel as a stated goal. The abundance of existing scientific research into the behaviors of materials in space meant that more meaningful contributions to knowledge could be made by highlighting other aspects of the work. Hope resisted pressure to declare that she would produce a liquid sculpture, and instead focused on the value of the experience of experimentation in the variable gravity environment, presenting the work as an exploration of play.

Despite having permission to approach the work as an artist, Hope found herself affected by the scientific environment that surrounded her on the flight. She describes "absorbing the scientific energy" from others who were conducting "serious" experiments alongside her, and found herself intensely focusing on carrying out her task as prescribed in her pre-flight plans. She was initially strapped to the cabin floor, in order that she could maintain a fixed posture and focus her attention on the task at hand. As she worked, however, she realized that this restraint was forcing her into an artificially scientific mindset, in which all of her actions and gestures were measured and controlled. She began to feel constrained by her situation and made a spur-of-the-moment decision to unstrap her legs and allow her body to float freely. This was a turning point in Hope's experience. Having made the decision to "go off-script," she discovered that her experience was transformed.

The difference between Hope's experiences of working in constrained and unconstrained positions was "mind breaking and transformative," and her relationship to the work was profoundly affected by this change. The video recording of Hope's experiment captures the contrast between her actions before and after she releases her legs. While her body is secured in place, her gestures are purposeful and controlled. She draws regular spirals, producing a consistent stream of blue glue. After her body is released, she appears to struggle to establish and maintain a consistent connection between the stream of glue and the wooden anchors. The fluid movement that was initially demonstrated by the liquid glue appears to be transferred to her body, as she floats in the space beyond the containment cube. The effort that she initially invested in controlling the flow of glue is now invested in controlling her body and its relationship to the experiment. The work becomes a whole-body performance.

In hindsight, Hope notes that this moment of rebellion was essential in achieving her aims. By releasing herself from the floor, the "messiness of craft" that is so important to her project could then be mirrored in her body, which, in its liberated state, was no longer fully under control. Hope appreciates that the need for strict adherence to a plan has arisen from a need for safety, "but also as an artist. I want room to play an experiment" (*sic*). Play, stresses Hope, is impossible unless there is some allowance for artists who feel the need to follow their intuition during an experiment. There needs to be some give-and-take, and compromises reached on both sides, where artists appreciate the importance of safety while spaceflight providers work with them to enable the unexpected. This lesson may benefit the spaceflight sector as well as artists with whom they engage, as preparation for a future when spaceflight participants increase in number and diversity. There can be "a tremendous amount of value in figuring out how to safely create the conditions for experimentation."

Hope's experiences remind us that, unless future space travelers are free to operate intuitively and make impulsive decisions, there will be untapped potential in many engagements with space. Activities that are carried out in space are strictly scripted to the extent that there are very few opportunities to deviate from an approved plan. Even where ODFs are not in use (see Chapter 2), an environment that encourages scientific approaches may be unintentionally inhibiting. Artists benefit from cultivating the unexpected, and expectations that are shaped by a history of science, create additional barriers to the production of meaningful art. Beyond the arts, the value of play is beginning to be recognized as essential to the psychological wellbeing of astronauts, and artists play a role in demonstrating how play can be encouraged in their strictly controlled environment (Balint and Lee 2019). *Space/Craft* contributes to this emerging appreciation for the importance of play in spaceflight. Hope asserts that "joy, play, experimentation [and] freedom" are "a core part of what it is to be human. For those things to be missing in any sort of future, whether on space or Earth, is deeply sad."

Alexis Hope was interviewed on March 19, 2023.

Case Study 9
Sana Sharma, *Fluid Expressions*, 2020–22

MIT designer and researcher Sana Sharma observes that, since paint behaves differently in microgravity, new tools need to be developed for its application. Sharma's *Fluid Expressions* explores "how the unique properties of fluids in microgravity can be used to create novel artistic tools." Sharma's project incorporates experiments into the behaviors of paint in microgravity, and takes a step further, conceiving new tools and mechanisms to exploit this behavior for artistic purposes. Her project began by asking how we might reconsider the brush and canvas for microgravity environments, and what new tools might be developed to paint in new ways (Sharma 2020). In place of a brush, Sharma's project utilizes sponges, rods, and rings, each of which can be used to manipulate free-floating drops of paint. These are housed within cylindrical canvases so that paint may be deposited in any direction. These were in-turn housed in a capsule that Sharma hand-produced from wood, acrylic, and leather. *Fluid Expressions* invites artists to set aside terrestrial tools, and makes progress toward redefining the practice of painting for the weightless environment.

Responding to previous microgravity painting experiments by Frank Pietronigro and Nicole Stott, and the calligraphy of JAXA astronaut Koichi Wakata, Sharma reflected on the extent to which terrestrial painting tools, specifically the brush and canvas, work differently in microgravity, where both the weightlessness of the artist and the behavior of paint in this environment affect the act of painting. She observed that astronauts typically take one of two approaches in their activities and experiments: either they employ terrestrial tools and make-do or manipulate those tools to account for the effects of microgravity, or they develop new, often makeshift, apparatus for the microgravity environment. In the act of painting, there are terrestrial tools and approaches, as in the traditional paintbrush and flat canvas employed by Nicole Stott in her watercolor painting (see page 151); and there are alternative approaches such as the one taken by Pietronigro (see Chapter 2), in which paint is squeezed into the air and allowed to float toward a three-dimensional canvas that surrounds the whole experiment. The tools employed by Stott and Pietronigro were, respectively, a paintbrush and pastry bag. The paintbrush, observes Sharma, features bristles that are redundant in microgravity, as water flows around them rather than being absorbed by them. The pastry bag, as employed by Pietronigro, enables an alternative approach to depositing paint, but was appropriated from a different terrestrial activity, cookery. Where these terrestrial tools do not function as they do in Earth gravity, Sharma concluded that there is scope for new apparatus that enables paint to be deposited on canvases in new ways.

The design of Sharma's painting capsules combines tools employed in astronauts' experiments and travelers' painting kits found on Earth. She transformed the flat canvas of Earth into cylindrical and spherical canvases that would be more fitting for a microgravity environment where multidirectional operation is the norm. These canvases were housed inside cylinders alongside an injection system that could be used to introduce paint into the canvas container. In an array of different cylinders, Sharma adapted the system to incorporate wet and dry paint, and different methods of depositing that paint onto canvases. The cylinder housing is constructed by hand from wood, acrylic, and leather (see Figure 6.6). This choice of materials was in-part inspired by travelers' painting kits that have historically been used on Earth. These kits are traditionally wooden boxes or leather cylinders or rolls that are used for the storage and transportation of canvases or paper, paint and brushes. Wood and leather take on additional significance during space travel, suggests Sharma, as although they are familiar on Earth, they are among "the rarest in the universe overall," and therefore give the design "strong terrestrial roots" (Sharma 2022).

Onboard an MIT-funded parabolic flight in 2021, Sharma deployed cylinders containing three different set-ups: first, a sphere of dried gouache and a wet canvas; second, free-floating spheres of wet paint and a dry canvas; and third, wet paint alongside a moveable rod and dry canvas (see Figure 6.7). Having observed Don Pettit's fluid experiments onboard the ISS, in which rods and rings are used to anchor water and ink globules and films, Sharma was aware of the potential for wet paint to form globules that would become attached, through surface tension, to metal rods and other

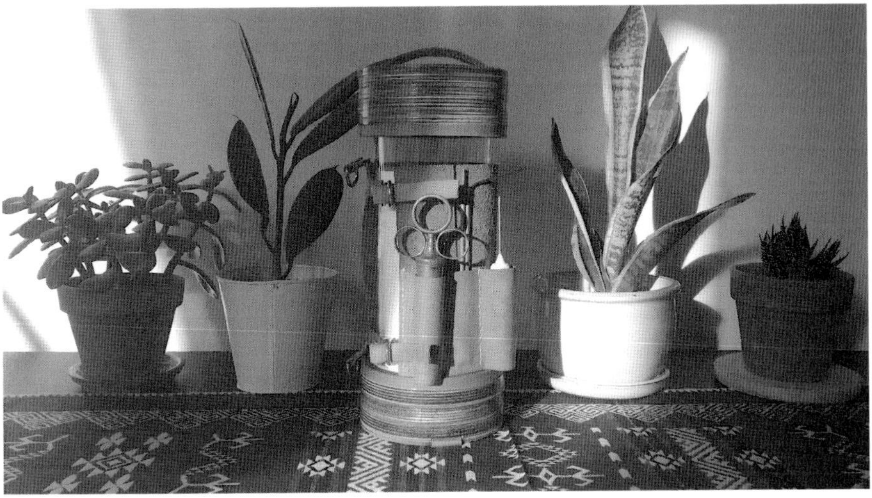

FIGURE 6.6 Sana Sharma, painting kit for *Fluid Expressions*, 2020–22. Sharma's cylindrical kit is handmade from terrestrial materials including wood and leather. The design is informed by historical traveling painter's kits. Image courtesy Sana Sharma.

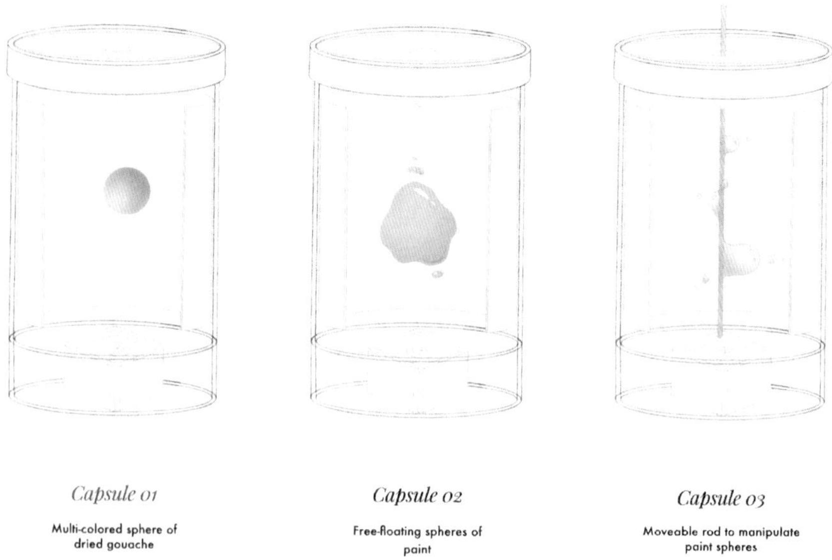

FIGURE 6.7 Sana Sharma, *Fluid Expressions*, 2020–22. Three capsules deployed onboard a parabolic flight in 2021. *Capsule 01* includes a sphere of dried gouache and wet canvas, so that the paint will mark the canvas when coming into contact with its wet surface as a result of motion in microgravity. *Capsule 02* includes a dry canvas. Watercolor paint is injected into the cylinder in small bursts, and the container is rotated to direct the paint deposition. *Capsule 03* includes a telescopic rod that is used to drag watercolor paint along the canvas. Image courtesy Sana Sharma.

surfaces. Onboard the flight, Sharma interacted with the three cylinders in ways that deposited abstract marks on the surrounding canvases (see Figures 6.8–6.9). The first cylinder was rotated so that the free-floating dried gouache would deposit marks on a surrounding wet canvas; in the second cylinder, paint was injected at intervals as the canvas was rotated; and in the third cylinder, the metal rod was used to collect floating globules of paint and deposit them dragging the paint across the canvas surface.

Sharma's first flight taught her about the behaviors of materials as well as how it is necessary to engage with microgravity as a kind of "co-creator." Weightlessness causes watercolor to become "sticky" due to surface tension, as if it has increased viscosity, attaching itself to any surface that it encounters, without necessarily being absorbed at the point of contact. Sharma found that when wet paint was injected into her capsule it would bounce across the canvas, leading to "running and smearing" that she had not predicted before the flight. These marks were very different to the marks made by the dry gouache on wet canvas, which had begun to dry by the time

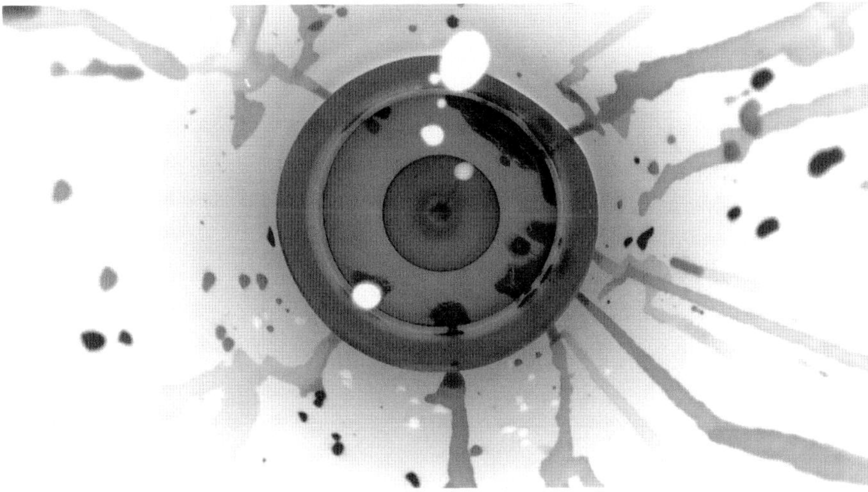

FIGURE 6.8 Interior view of *Fluid Expressions*, 2020–22. Paint is deposited inside *Capsule 02* during a period of weightlessness onboard a parabolic flight. Image courtesy Sana Sharma.

the plane entered its first parabola, leading to hard-contoured marks that resemble those made by dry pastels.

For a later flight in 2022, Sharma developed a larger cylinder that could contain a spherical canvas (see Figure 6.10). This version drew from Sharma's own roots, employing an Indian tie-dye technique, Bandhani, in which thread is wound around pinched fabric before dying. Silk, chosen for its greater absorbency than Sharma's previous canvases, was wound around a hollow ball. The ball was then allowed to float freely inside the large cylinder as dye was injected through the lid of the cylinder. The whole cylinder was strapped to Sharma's body like an accordion, with the cylinder angled along the transverse plane of her body, and syringes on its lid positioned like accordion keys. In this position she was able to "play" the capsule, pressing the keys (syringe plungers) to release ink into the chamber. Sharma views this second iteration as the product of an agathonic design process, in which the design itself has evolved, and she, as its user, has further adapted to using it in the weightless environment.

In this second iteration, Sharma's body plays a significant role in the dying of her silk canvas. There is no direct method of depositing the dye on the silk; rather, the dye is released into the cylinder and then the spherical canvas is moved into a collision course with the floating dye through a combination of weightlessness and the movements of Sharma's body. In this arrangement, the cylinder became "an extension" of Sharma's body, so that her body and the cylinder functioned as part of the same system, the movement of one affecting the movement of the other. This two-body

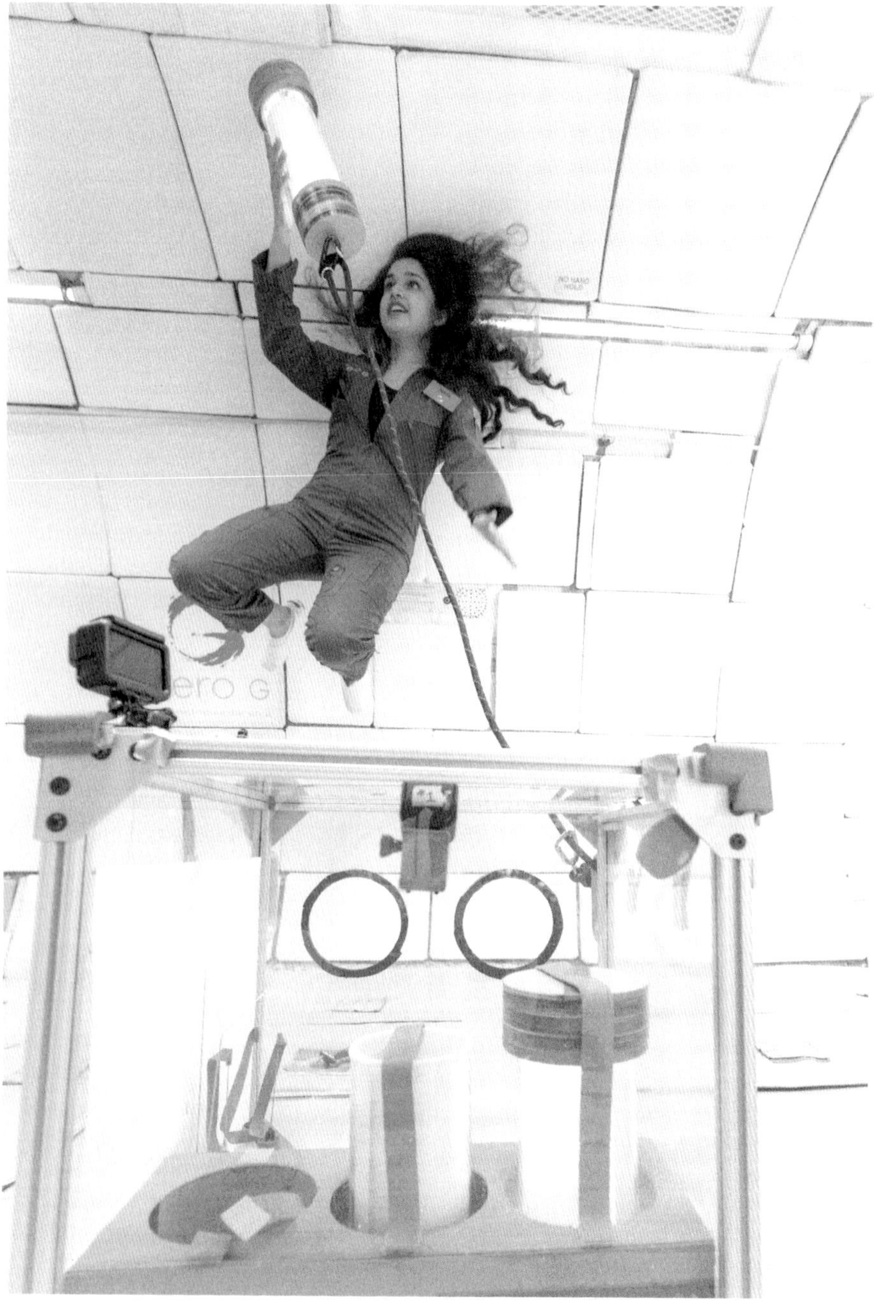

FIGURE 6.9 Sana Sharma deploys *Fluid Expressions* onboard a parabolic flight in 2021. Sharma rotates a cylinder to direct a paint deposit onto a cylindrical canvas. Image courtesy Sana Sharma, © Steve *Boxhall/Zero-G*.

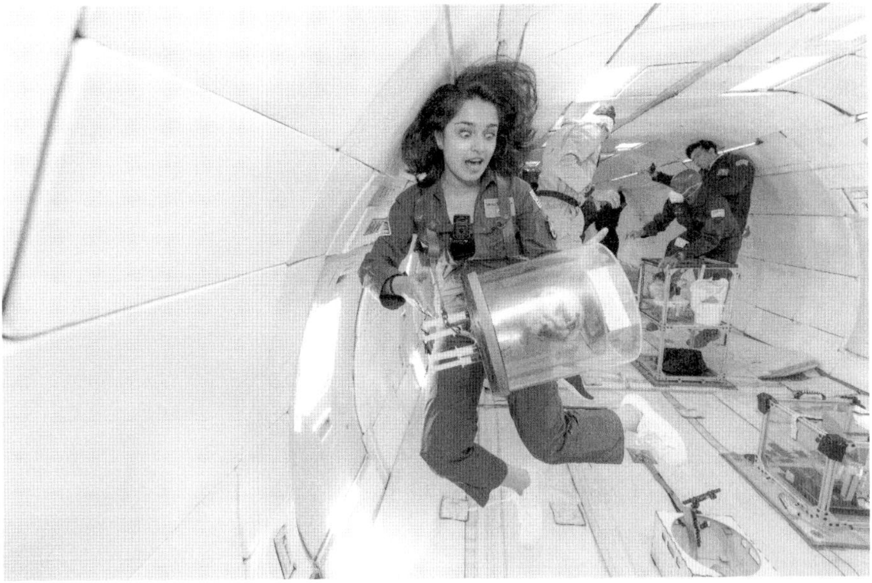

FIGURE 6.10 Sana Sharma operates a cylinder containing a spherical canvas onboard a parabolic flight for *Fluid Expressions* in 2022. The cylinder and Sharma's body become a two-body system, spinning around a center of gravity that is located in the space between both bodies. Image courtesy Sana Sharma, © Steve *Boxhall/ Zero-G*.

system had a center of gravity outside of Sharma's own body, so that her body and the cylinder were "both spinning around this central locus."

Sharma's work challenges established ideas about the roles that the body, brush, and canvas play in mark-making. Her experiments ask us to revisit the idea that paint is deposited on canvas, and imagine the reverse, in which the canvas moves to make contact with paint. Weightless performers, such as those cited in Chapter 5, report an awareness that their movement and location is always relative to surrounding objects and surfaces. Sharma's second iteration of *Fluid Expressions* extends this experience to productive interactions with other objects, leading to a process of mark-making that sets both the canvas and the paint on a collision course with one another. In this system, neither the location of the paint nor the canvas can be taken for granted, and the depositing of the paint marks the moment at which both media collide. Painting in this way thereby becomes less about deposition than it is about alignment.[1]

[1]Parallels can be drawn with Earth-based practices that involve placing canvas into media, such as marbling, the practice that inspired Takura Osaka's *Marbling Painting on a Water Ball*, in which paper sheets are brought together around a weightless globule of ink, so that the ink transfers onto the paper (see page 159). However, in these established, terrestrial practices and the space art that they have inspired, the fluid medium is static relative to its surroundings, and the artist is responsible for moving the canvas into contact with the medium.

In Sharma's "chaotic" system, the artist has very little agency. The body, the paint, and the canvas are all afloat. The result is, at least in these experimental phases, more a work of serendipity than of humanmade production. Sharma contrasts her own experience with those of more experienced space travelers who might be able to exert more control over the relative movements of the cylinder and their body, and therefore be more able to purposefully compose an output. As the designer of a new set of tools, she is not yet an expert in their use, and will depend on future users to test their potential. As with any artists' tool, the *Fluid Expressions* systems will be deployed in different ways in different hands, resulting in different styles of output. Having demonstrated their potential for use in as diverse practices as watercolor painting and Bandhani dyeing, Sharma has illustrated the potential for a wide variety of art and craft practitioners to develop their practice using her "ways of making."

Sana Sharma was interviewed on January 12, 2024.

Conclusion: Finding Space for Artists

Space exploration is as much about the discovery of new phenomena as it is about the discovery of new territory. Research takes place in Earth orbit because humankind still has much to learn about the effects of weightlessness, and humankind continues to be awe-inspired by the results of experiments that seem to challenge the models and behaviors that we take for granted in Earth gravity. The arts are lagging behind science and engineering in their engagement with this novel environment, but there is growing recognition of the potential for the arts to contribute meaningfully to our future in space, and of the ways in which space may transform art practice. The new phenomena discovered as a result of space exploration lead to new tactile and visual experiences that can be harnessed in new art practices.

There are still many barriers that artists must overcome in order for them to be able to demonstrate the potential and value of art in orbit, but opportunities are increasing. As the space sector expands and access to space is democratized, an increasing variety of opinions as to what constitutes value in space engagement inform a widening range of initiatives. As these barriers are broken down, artists have increasing agency over the kinds of engagements that they may propose for space, and the way that spaceworks are presented to audiences. It is up to artists to seize these opportunities when they arise, and to use them in responsible ways. This book argues that space artists should take responsibility for demonstrating the potential for a new field of practice and to discover, and then communicate to audiences, what new practices are possible in the environment of space.

The examples in this book contribute to demonstrating the potential for space as a site for art practice, and highlight some of the new practices that this environment makes possible. They demonstrate that new kinds of work can be produced in space that could not exist in a terrestrial environment, that new effects can be produced, and that profound new experiences can shape the way that we view ourselves and our relationship to our environment. They also demonstrate that there can be a variety of different approaches to art in orbit; that orbital art is not one thing, but many. Art in orbit can sometimes be intended to last for eternity, and at other times is intended to be finite, burning up in Earth's atmosphere as it de-orbits. While artists including Eduardo Kac imagine that their works might continue to float through space indefinitely, to one day be discovered by some unknown future civilization, conversely, other artists have set out to create work that will, like Peter Beck's *Humanity Star* or Azuma Makoto's *Exobotanica*, have a limited lifespan, sometimes showing that destruction becomes a meaningful part of a spacework's journey. Some artists have chosen to produce spaceworks that explore the novel behaviors of familiar materials in microgravity, as when liquids become sculptural in Meroz and Segal's *Impossible Object* or Alexis Hope's *Space/Craft*. Others, like Eyal Gever or Arthur Woods, have produced works that are more solid and permanent but that are experienced as transient as they revolve and float through space.

Many of the artists interviewed for this book found it important not to speak about their spaceworks in isolation, but rather to locate them within a "story." "Art is always bigger than all of us. It has drama, it has a story," says Eyal Gever. These stories describe the artists' experiences of navigating their relationships with the spaceflight industry and with the astronauts who deployed and interacted with their work; the experiments that preceded the deployment and the lessons learnt when witnessing the work finally deployed in microgravity; and the process and methods of recording and how these recordings reached audiences on Earth, including the artists themselves.

One of the questions arising from the examples in this book is, how do Earth-bound audiences see or experience art in orbit? Trevor Paglen's *The Last Pictures* and Max Baraitser Smith's *Lunar Bureau of Weights and Measures* are intended to provoke conversation among audiences who will never see the actual artefacts for themselves. In these cases, the story surrounding the work may have more impact than the work itself. Where a spacework is represented on Earth, either through duplication, recordings, or exhibition of the spacework itself following return to Earth, questions arise about whether the viewing of these representations constitutes seeing and/or experiencing the work, given that these viewings do not take place in the environment for which the work was intended. Can we consider these showcases to be exhibitions of the work, or are they another part of the story? Jeanne Morel attests that visual representations of the way that objects or bodies behave when weightless are not sufficient to help audiences understand the effects of weightlessness. Projects like those of Morel and Marlier prompt consideration of what it means to *see* the effects of weightlessness. Jeanne Morel's microgravity dances are recorded as data and then reinterpreted as visuals that are not direct visual representations of the dance itself but of the data that Morel's body and mind generate during her performance. Are these representations sufficient to demonstrate the conditions of space or the value of engaging with those conditions? Do they compromise the integrity of art in orbit, transforming it into the kind of astroart that visualizes or depicts space without the need to engage directly with that environment? Are Earth-bound representations tokenistic in ways that undermine the goals of artists who seek new ways to engage with space? These questions are important for anyone who seeks to assert the value of art in space and make a case for support for future endeavors.

Interviews with artists, recorded as this book's case studies, have revealed the ways that engagement with space has profoundly affected their approaches to practice, but also recorded the difficulties and constraints that shaped their projects. They demonstrate the extent to which artists hoping to produce work for space must compromise or limit their work in the current climate. Reliance on small start-ups, emerging businesses, and philanthropists means that artists who rely on them are subject to restrictions that limit the size and scope of their work. Where larger institutions or government organizations enable space art, their agendas can be narrow or out of alignment with the needs of the art sector. Many of the artists featured in this book have had to work at a very small scale (less than 1 cm^2 in several cases); were limited in their choice of materials; have had to rely on others to produce or deploy their spaceworks on their behalf; have had to reframe their proposals as science experiments; and have had no opportunity to test their work before final deployment.

As is often the case when access is controlled by gatekeepers, new practice must be presented in terms that make sense to, and meet the approval of, stakeholders from other disciplines. Access to space is controlled by the space sector, and inevitably this means that ideas about the nature and value of space art have been shaped by scientists and others, often without a full understanding of the arts. It is the responsibility of the first generation of space artists to ensure that human's vision of art in space is not dictated to them by other disciplines, or by initiatives that do not fully recognize the potential for innovation in this new field. It is also the responsibility of stakeholders from other fields to recognize that they must involve professional artists in conversations about what the arts can contribute to their activities. The space sector must maintain awareness of the power that they have and the ways that they may exercise it.

If the space sector is willing to allow artists to become more involved in spaceflight in the future, they must commit wholeheartedly, supporting the whole arts sector, not just individual artists or projects. Frank Pietronigro (2004) cautions against allowing one artist to represent the whole arts sector, and thereby define space art in their own terms:

> I believe it would be a mistake to send one individual as the "first artist in space" upon the moment that space agencies first decide to send artists into space. I strongly advocate that an international team of artists should

> make this significant journey together … I believe that no individual artist, nor their chosen media or creative process can represent the significant development of using space flight technology for artistic production.

Some artists in a privileged position of being able to create opportunities for their peers have recognized that they have a responsibility to do so. In the spirit of "participation and sharing," after having launched his own work into space (see Chapter 1), Xu Bing, founded the *Satellite Creative Residency Project* in collaboration with the Beijing Wanhu Chuangshi Cultural Media Co. (Bing 2024). In 2024, Bing's SCA-1 dedicated arts satellite provided space for project residents to launch their own work into orbit. The initiative seeks to expand and diversify the field of space art, and in particular encourages young artists whose subsequent careers might then be influenced by their formative experiences of producing work for space. Similarly, Yusaku Maezawa's dearMoon mission recognizes the importance of populating any space arts mission with a variety of practitioners from different backgrounds and disciplines. The mission is set to include artists working in a range of contemporary media and professions.

When I originally proposed this book, my intention was to include dearMoon as one of its case studies. The dearMoon mission was, at the time, tentatively scheduled for launch toward lunar orbit in 2023. Since then, however, it has been placed on hold pending the completion of the SpaceX Starship, which is currently still in testing phase. dearMoon is just one of many planned missions and projects that have been delayed, substantially redesigned, or withdrawn altogether during the writing of this book. The space sector grapples with uncertainty of all kinds, being so reliant on the development of emerging technologies as well as the political environments that may prioritize or deprioritize space exploration at any moment. The interviewed artists have found that, the more ambitious the mission, the more likely it is to be postponed or canceled, and this risk is even greater for arts projects that are low on the list of priorities for spaceflight providers. It is my hope that, by the time this book is published, there will be more spaceworks on the way to space, and more artists recognizing the potential for their own work to join the examples recorded in this text.

In the introduction to this book, I noted that we are only just beginning to engage with space, to develop human space culture, and we still, observes philosopher Elie During (2019), understand spaceflight "according to … earthly frameworks." There is still a long journey of space culturalization ahead of us, beginning with activities that transport terrestrial culture into space, and perhaps one day resulting in culture that emerges solely from space. Spaceworks can help us to begin to realize a distinction between *human culture* and *terrestrial culture*, moving toward a future in which space culture does not always look back to Earth, where terrestrial conditions do not limit our imagination.

In order to achieve this, we must invest in space art that explores the potential of the environment of space, not art that reinforces terrestrial ideals and practices. While I have been writing this book, six time capsules, collectively known as the *Lunar Codex*, have been landed on the lunar surface (Schrader 2023). This initiative recorded the work of 35,000 artists on a series of NanoFiche discs. Each disc has been enclosed in a capsule and deposited at one of six locations on the Moon. The collection of works represents the Earth-bound human, and the Earth itself. Representative images overwhelmingly favor depictions of nature (including landscapes and animals) and domestic environments, as well as human figures. It is also important to highlight that the human subjects of these representative images are predominantly female, in contrast to the male-dominated history of space exploration (http://www.lunarcodex.com). To this extent, the project reinforces notions that Earth is the domestic, homely environment of the female, while space is the domain of the heroic, adventuring male. It is important to recognize that such projects are not representative of space art. The purpose that they serve is to archive particular visions of the Earth. In other words, they may extend terrestrial culture, but do not show us the full potential of human culture when expanded to new environments beyond the Earth.

This book has argued that art in space is a new field that will have distinct disciplinary practices. Annick Bureaud (2003) suggests that the distinction between terrestrial and space art is not so straightforward. Writing about works for microgravity, Bureaud suggests that these works "have not invented a new 'genre.' They have inscribed themselves into well-established disciplines such as dance [and] sculpture." These works "build on what we know," "bringing about a new language" for the existing disciplines of dance, painting or sculpture. If space art is a new "language," then this language is transdisciplinary. The existing terrestrial art disciplines continue to exist in space, but they collectively adopt new concerns and each respond to those concerns in discipline-specific or cross-disciplinary ways. Some disciplinary boundaries are challenged by microgravity, as when painting becomes sculptural, and hence it is not always possible to trace a clear path from terrestrial practices to space art practices within existing disciplinary boundaries.

So, we should not forget our Earthly origins. This book has aimed to demonstrate that art in space can move forwards, away from Earth, embracing the new possibilities that are provided by spaceflight, but Earth will always be the birthplace of human culture. It is possible to acknowledge and respond to our relationship with Earth without being limited by terrestrial thinking. Thematically, spaceworks may represent aspects of the Earth while also speaking about alternative modes of existence. New spaceworks can not only remind us that we are human, but also that what it means to be human can change.

It is also the case that art in space can benefit audiences on Earth. Just as the science experiments onboard the ISS can yield discoveries that improve life on Earth, working in space can yield new ideas that can be manifested in work in Earth as well as in space. Artists who have experienced weightlessness have described a postgravity experience that leads them to think differently about gravity on Earth, and hence to work in new ways. The experience of spaceflight can empower artists to think more critically about the practices that have developed in our terrestrial environment, and then to develop new ideas and new practices that respond to the way that the terrestrial environment is unique in the universe. In this way, spaceflight empowers artists to better respond to our Earthling condition and the uniqueness of our home planet.

While this book has made the distinction between Earth and space, this binary reflects the current state of space exploration. Over the rest of the twenty-first century, humankind will make further progress in the colonization of space. We will extend our reach further into open space, and on to the surface of other planets. We will encounter a greater variety of spaces, each of which may become a site for other new kinds of creative practice. Max Baraitser Smith's proposal for the *Lunar Bureau of Weights and Measures* shows us that work on other planets might need to exist within new systems and structures, or may inform those new systems and structures as they emerge. Lisa Pettibone suggests that, in these new environments, there may be "new kinds of art we haven't even thought of yet." Several of the artists that were interviewed for this book expressed interest in further creative engagements with space, including the opportunities that might exist on the Moon or Mars. Pettibone and Eyal Gever both expressed excitement about the possibility of making use of the materials available at extra-terrestrial locations. Artists have always made use of materials in their environment, and the Moon and Mars will provide new other-Earthly materials for creative practice. Pettibone speculated that clay might be extracted from the lunar surface, and Gever described the potential for Moondust to be the raw material for three-dimensional printing, hardened with a beam of light.

As an educator, I am concerned with the question of how, and why, to teach space art. Art education is so often about tactile and experiential learning that it benefits from taking place in the conditions and locations of art production, as in the workshop or studio. Few educational institutions have been able to offer their students hands-on experience of weightlessness—MIT being the main exception—and those opportunities are mainly reserved for researchers rather than students. In my own teaching I have used water tanks and trampolines to replicate, albeit in flawed ways, the microgravity environment. These experiments have given students some limited understanding of the novel forms that they can create when materials are not so weighed down by gravity but, ultimately, needed to be supplemented by Zoom calls with astronauts who were able to offer verbal accounts of their experiences of weightlessness. In order to teach space art it is currently necessary to resort to indirect observation, through secondary research and thought experiments. These methods may sometimes be combined with practical experiments that are designed to prompt students to more consciously consider the effects of gravity. In this compromise, the value of teaching space art is revealed. While it may be true that most art students of this generation will never produce space art, the benefits of teaching space art practice can be seen in their work for Earth. Prompting consideration of the extent to which Earth gravity affects their existing practice, students are taught to take a more critical and experimental approach to some key aspects of their practice, and to be more adventurous in their approaches to orientation and weight. Students who took part in space art and space design projects went on to be more experimental in their subsequent work, particularly in their approaches to weight, balance, and form.

Since I began writing this book, the commercial space exploration landscape has evolved and expanded, meaning an exponential increase in the number and diversity of opportunities for artists and art students to send works into orbit. Small businesses routinely send satellites into orbit, and make spaces available to customers at increasingly affordable prices. The cost of deploying a spacework does not necessarily equate to the cost of launching an object into space. As Eduardo Kac demonstrated with his *Télescope Intérieur* (see Case Study 6), art can be sent into space in the form of an idea or instructions, and produced using materials that are already in space. Objects that might otherwise be considered space junk can be repurposed after their functional life has ended, as per the example set by Xu Bing. Bing's *Satellite Lake: Cosmic Reflections* made use of Commsat's *Ladybug-1* satellite after it had been decommissioned (see Chapter 1).

In these ways, the price of deploying work in space is gradually becoming comparable to the cost of exhibiting in galleries on Earth. While writing, I found it surprisingly straightforward to identify a commercial company willing to accept art payloads onboard a cubesat to be launched into low-earth orbit. UK-based AmbaSat, who provide satellite kits and associated educational materials to education providers, were able to accept art payloads from students at the Royal College of Art. We invited students to propose two kinds of payload: an art object that would fit AmbaSat's 35 × 35 mm standard flight tray; and an Earth-based live-data visualization that would receive and visualize data from sensors onboard the satellite during its month in orbit. The selected works combine both of these, using sensors and circuit-boards that could gather and transmit data as the canvas for micro-etched artworks. At the time of writing, these works are scheduled to be launched into orbit in 2025.

One lesson that this project has taught us is that spaceflight is now sufficiently advanced that there are fewer unknowns than there have been for some other artists' previous works and experiments. The RCA/AmbaSat project did require our students to operate with strict constraints, including the small size of the flight tray, but as these constraints were known at the time of the proposal they were not detrimental to the project. The two most significant factors that informed the meaning and significance of the proposed works were the fact that the work could not be directly viewed while in space, and that it would burn up during de-orbit. The fact that the work could not be viewed directly helped to ensure that the work would not be tokenistic, as proposals would be required to exhibit some greater purpose or significance beyond aesthetics. The fact of it being in orbit would need to be meaningful beyond simply or superficially generating images of the work against the backdrop of outer space. This factor also prompted conversations about how the satellite's journey could be visualized in other ways, including through interpretations of the live data that would be beamed back to Earth throughout its month in space. Knowing that the satellite would eventually burn up, students proposed works that would bring meaning to their destruction.

I will conclude this book with an invitation to astronauts to play a role in the expansion of the field of space art, in the form of an Art Operations Data File (Art ODF) that can be performed by anyone traveling onboard a spaceship without any specialist art tools or equipment:

1. Onboard a spaceship, float weightlessly, untethered to your surroundings. Feel the void.
2. Turn to the nearest window and gaze at the distant Earth. Experience the overview effect.
3. Close your eyes. Imagine art in space.

APPENDIX

TABLE 1 Timeline of cited spaceworks

Launch	Title	Artist	Site
1971	*Fallen Astronaut*	Paul Van Hoeydonck	Moon
1975*	*Beacon*	Albert Notarbartolo	Open space, Earth orbit
1975*	*Star Cloud*	Albert Notarbartolo	Open space, Earth orbit
1986*	*ARSAT*	Pierre Comte	Open space, Earth orbit
1986*	*L'Anneau de Lumiére*	Group Spirale	Open space, Earth orbit
1987*	*Pacific Triangle*	Pierre Comte	Earth (to be viewed from space)
1988	*Drift Painting*	Frank Pietronigro	Parabolic aircraft
1993	*Cosmic Dancer*	Arthur Woods	Mir Space Station, Earth Orbit
1993	*Space Atelier*	Niu Bo	Parabolic aircraft
1994	*Gravity Zero*	Kitsou Dubois	Parabolic aircraft
1995	*Noordung 1995:2045*	Cosmokinetial Kabinet Noordung	Parabolic aircraft
1997**	*Monogram*	Eduardo Kac	Deep space rocket, Saturn orbit
1999	*Trajectoire Fluide*	Kitsou Dubois	Parabolic aircraft
2001	*Falling Without Fear*	Morag Wightman	Parabolic aircraft
2001	*Wave/Particle*	Ansuma Biswas and Jem Finer	Parabolic aircraft
2001	*Zero Genie*	Ansuman Biswas and Jem Finer	Parabolic aircraft
2002	*Dansez Maintenant*	Renaud Auguste-Dormeuil	Parabolic aircraft
2003	*Constellation Columbia*	i-Dat	Parabolic aircraft
2003	*Dédalus Project*	Marcel.li Antúnez Roca	Parabolic aircraft
2003	*Kapelgraph Zero-G*	Vadim Fishkin	Parabolic aircraft

Launch	Title	Artist	Site
2003	*Sound Wave Sculpture*	Takuro Osaka	Parabolic aircraft
2003	*Spot Painting*	Damien Hirst	Mars
2004	*Hiten (Flying Deities)*	Setsuko Ishiguro	Parabolic aircraft and ISS Earth orbit
2007	*Marbling Painting on a Water Ball*	Takura Osaka	ISS, Earth orbit
2008	*Artistic Experiments Using a Water Sphere*	Takao Fujiwara	ISS, Earth orbit
2008	*Homage to Francis Bacon: Triptych I* and *II*	Nasser Azam	Parabolic aircraft
2008	*Singular Oscillations*	Bradley Pitts	Parabolic aircraft
2009**	*Lagoogleglyphs*	Eduardo Kac	Earth (to be viewed via satellite)
2009	*Perspectives*	Kitsou Dubois	Parabolic aircraft
2009**	*Spiral Top*	Takuro Osaka	ISS, Earth orbit
2010*	*Nomadic Nature Kit*	Kirsten Johannsen	Deep space rocket
2011	*Light, Life and the Sea*	Takura Osaka	ISS, Earth orbit
2013	*The Last Pictures*	Trevor Paglen	Communications satellite, Earth orbit
2014	*Exobotanica I*	Azuma Makoto	Stratosphere
2014	*Holding Air*	Nahum	Parabolic aircraft
2014	*Un infinito*	Ale de la Puente	Parabolic aircraft
2016**	*Danse en zerogravité*	Jeanne Morel	Parabolic aircraft
2017	*Orbit Weaver*	Xin Liu and Andrea Lauer	Parabolic aircraft
2017	*Télescope Intérieur*	Eduardo Kac	ISS, Earth orbit
2017	*Spiral Top II*	Takuro Osaka	ISS, Earth orbit
2017	*#Laugh*	Eyal Gever	ISS, Earth orbit
2018	*La Chaise*	Kitsou Dubois	Earth
2018	*Humanity Star*	Peter Beck	Open space, Earth orbit
2018	*Orbital Reflector*	Trevor Paglen	Open space, Earth orbit
2018	*SpaceHuman*	Velentina Sumini	Parabolic aircraft
2019	*Long Distance Low Gravity Love Letter*	Ani Liu	Suborbital rocket

Launch	Title	Artist	Site
2020	*Abiogenetic Triptych*	Andrea Ling	ISS, Earth orbit
2020	*Memory Chain: A Pas de Deux of Artifact*	Kat Kohl	ISS, Earth orbit
2020	*Nothing, Something, Everything*	Masahito Ono	ISS, Earth orbit
2021**	*Fluid Expressions*	Sana Sharma	Parabolic aircraft
2021	*Human Frontiers*	Arti Struyanskiy	ISS, Earth orbit
2021	*Lunar Bureau of Weights and Measures*	Max Baraitser Smith	ISS, Earth orbit
2021	*Satellite Lake: Cosmic Reflections*	Xu Bing	Satellite, Earth orbit
2022**	*Adsum*	Eduardo Kac	ISS, Earth orbit (later to be landed on the Moon)
2022	*Impossible Object*	Yasmine Meroz and Liat Segal	ISS, Earth orbit
2022	*Nothing In Space*	Ohad Fishov	ISS, Earth orbit
2022	*Space Jewel*	Freddy Kislev	ISS, Earth orbit
2022	*Verdant*	Lisa Pettibone	ISS, Earth orbit
2024**	*Ágora*	Eduardo Kac	Deep space rocket, Transplanetary orbit
2024	*Moon Phases*	Jeff Koons	Moon

*Proposed or unrealized work.
**Launch date indicates year of first launch, where work has continued into subsequent years through further deployments, through deployment of further iterations, or is deployed indefinitely onboard long-distance flights.

BIBLIOGRAPHY

Ackermann, E. (2016), "Microgravity Playscapes: Play in Long-Term Space Missions," *American Journal of Play*, 8 (2): 157–77.

Ahmadi, M. (2020), "The Experience of Movement in Orbital Space Architecture: A Narrative of Weightlessness," *Cogent Arts & Humanities*, 7 (1).

Arnheim, R. (1960), *Art and Visual Perception: A Psychology of the Creative Eye*, third printing, Berkeley: University of California Press.

AMKK (2014), "In Bloom #1: EXOBIOTANICA I-BOTANICAL SPACE FLIGHT," *Azuma Makoto*. Available online: https://azumamakoto.com/en/project/0r970tzezt45/ (accessed December 11, 2023)

Ars Electronica (2020), "Space has Arrived: Digital Exhibition: Sojourner 2020." Available online: https://ars.electronica.art/keplersgardens/en/digital-exhibition-sojourner/ (accessed March 21, 2024)

Ashcroft, F. (2000), *Life at the Extremes: The Science of Survival*, Berkeley: University of California Press.

Ashford, D. (2013), *Space Exploration: All That Matters*, London: Hachette.

Baba, A.I., M.Y. Mir, R.R. Riyazuddin, A.A. Cséplő, G. Rigó and A. Fehér (2022), "Plants in Microgravity: Molecular and Technological Perspectives," *International Journal of Molecular Science*, 23 (18): 10548.

Badovinac, Z. (2016), "Foreword," in B. Groys (ed.), *8th Triennial of Contemporary Art U3: Beyond the Globe*, 4–5, Ljubljana: Sternberg Press.

Balint, T.S. and H.L. Chang (2019), "Pillow Talk—Curating Delight for Astronauts," *Acta Astronautica*, 159: 228–37.

Balint, T.S. and C.H. Lee (2019), "Pillow Talk—Curating Delight for Astronauts," *Acta Astronautica*, 159: 228–37.

Bandla, S. (2023), "The Importance of Human Tended Experiments," paper presented at *2023 Next-Generation Suborbital Researchers Conference*, Broomfield, Colorado, February 27 to March 1.

Battaglia, D. (2012), "Coming in at an Unusual Angle: Exo-Surprise and the Fieldworking Cosmonaut," *Anthropological Quarterly*, 85 (4): 1089–106.

BBC (2021), "Japanese Billionaire Seeks Eight People to Fly to Moon," *BBC News*, March 3. Available online: https://www.bbc.co.uk/news/world-asia-56261574 (accessed January 6, 2023)

Beletsky, V.V. (2001), *Essays on the Motion of Celestial Bodies*, A. Iacob trans, Basel: Springer Basel AG.

Bell, L. (1983), "Design of Microgravity Space Environments to Enhance Crew Health, Morale, and Productivity," *The Space Congress Proceedings*, 5.

Benson, M. (1999), "Noordung Zero Gravity Biomechanical Theater," *NSK State*, December 22. Available online: https://nskstate.com/article/noordung-zero-gravity-biomechanical-theater/ (accessed March 21, 2024)

Benson, E., H. Kim, S. Jarvis, J. Norcross and S. Rajulu (2019), "Estimation of Preferred Work Envelope from Simulated Functional Tasks in a Spacesuit," Document ID: 20190028717, *National Aeronautics and Space Administration*. Available online: https://ntrs.nasa.gov/api/citations/20190028717/downloads/20190028717.pdf (accessed January 19, 2023)

Berengueres, J.M., M. AlKuwaiti, M. Abduljabbar and F. Taher (2023), "Adding Sound Transparency to a Spacesuit: Effect on Cognitive Performance in Females," *IEEE Open Journal of Engineering in Medicine and Biology*, 4: 190–4.

Berger, E. (2024), "Starlab—With Half the Volume of the ISS—Will Fit Inside Starship's Payload Bay," *Ars Technica*, February 1. Available online: https://arstechnica.com/space/2024/02/by-launching-on-starship-the-starlab-station-can-get-to-orbit-in-one-piece/ (accessed March 17, 2024)

Bertok, G. (2016), "On the Cosmos and Other Things," in B. Groys (ed.), K. Zakrajsek (trans.), *8th Triennial of Contemporary Art U3: Beyond the Globe*, 31, Ljubljana: Sternberg Press.

Bing, X. (2022), *Artworks*. Available online: https://www.xubing.com/en/work/details/691?classID=8&type=class#691 (accessed September 16, 2023)

Bing, X. (2024), "'Xu Bing: Art Satellite—The First Animated Film Shot in Space' is on View at Santa Veneranda in Venice. Central Academy of Fine Arts," *CAFA Art Info*, April 22. Available online: https://www.cafa.com.cn/en/news/details/8332457 (accessed September 13, 2024)

Biswas, A. and J. Finer (2005), "Zero Genie, Film, 2001," in N. Triscott and R. La Frenais (eds.), *Zero Gravity: A Cultural Users Guide*, 56–9, London: The Arts Catalyst.

Biswas, A. and J. Finer (2005a), "Wave/Particle," in N. Triscott and R. La Frenais (eds.), *Zero Gravity: A Cultural Users Guide*, 52–5, London: The Arts Catalyst.

Bizzarri, M., M. Monic and J.J. van Loon (2015), "How Microgravity Affects the Biology of Living Systems," *Biomed Research International*, 2015: 863075. Available online: https://www.ncbi.nlm.nih.gov/pmc/articles/PMC4312564/ (accessed March 8, 2024)

Brandt, P.A. (2013), "What is this Thing Called Space? Art and Spatial Imagination," *SSRN*, November 27.

Brownie, B. (2019), *Spacewear: Weightlessness and the Final Frontier of Fashion*, London: Bloomsbury.

Buckminster, F.R. and J. McHale (1965), *World Design Science Decade 1965–1975: Phase 1* (Document 4). Carbondale: World Resources Inventory.

Bureaud, A. (2003), "Art and Zero G. The Reasons for a Symposium," *Leonardo/Olats*. Available online: https://www.olats.org/annick-bureaud-visibility-legibility-of-space-art-art-and-zero-g-the-reasons-for-a-symposium-2003/ (accessed February 25, 2024)

Bureaud, A. (2004), "A New Territory of (In)human Proportions," *Art Press*, 298. Available online: https://www.annickbureaud.net/wp-content/uploads/2011/01/ArtPress-Spaceart.pdf (accessed March 8, 2024)

Bureaud, A. (2009), "Kitsou Dubois and the Weightless Body," *IEEE MultiMedia*, 16: 4–7.

Bureaud, A. (2021), "It's a Beautiful Name for a Satellite: Paradoxical Art Objects Somewhere between Politics and Poetics," *Leonardo*, 54 (1): 79–91.

Bureaud, A. and K. Dubois (2005), "The Embodiment of (Micro)Gravity. Kitsou Dubois's Analogies: An Artistic and Aesthetic Experience," *Leonardo/Olats*. Available online: http://archive.olats.org/space/13avril/2005-2/te_kDuboisBureaud.php (accessed December 30, 2023)

Carroll, M.W. (1982), "Space Art: The Impact of Space-Age Technology on Representational Art," *Leonardo*, 15 (3): 210–12.

Chang, Y.W.E. (2022), "Space-age Aesthetics in Modern Art Inspired by Space Exploration: A Brief Study on Installation, Painting and Music Art," *Acta Astronautica*, 199: 355–63.

Chao, D.F., R.D. Green, T. Hatch, J.B. McQuillen, W.V. Meyer, H. Nahra, P. Tin and B.J. Motil (2020), "*A Researcher's Guide to international Space Station Fluid Physics*," National Aeronautics and Space Administration, revised edition. Available online: https://www.nasa.gov/wp-content/uploads/2019/10/iss-fluid_physics_tagged.pdf (accessed December 26, 2023)

CNES (2022a), "Résidence ZERO-G," L'Observatoire de l'Espace du CNES. Available online: https://cnes-observatoire.fr/ode-dossiers-residences-zero-g.html (accessed September 16, 2023)

CNES (2022b), "Programme Avant-Poste Appel à projets de création," L'Observatoire de l'Espace du CNES. Available online: https://cnes-observatoire.fr/ode-projets/2022-02-15_residence-zero-g-2022/appel-a-projet-artistique-zero-g-2022.pdf (accessed March 21, 2024)

CNES (2023a), "L'exposition Avec l'Esoace, Vol. 2," L'Observatoire de l'Espace du CNES. Available online: https://cnes-observatoire.fr/ode-dossiers-expositions/2023_dossiers-avec-espace-vol-2/1_avec-espace-2_auguste-dormeuil.html#dormeuil (accessed September 16, 2023)

CNES (2023b), "Programme Avant-Poste: L'extase appel à projets de création en impesanteu," L'Observatoire de l'Espace du CNES. Available online: https://cnes-observatoire.fr/ode-projets/2023-08-30_programme-avant-poste-zerog-2024/appel-a-projet-artistique-zerog_2024.pdf (accessed September 16, 2023)

CNES (2024a), "Programme Avant-Poste: Appel à Manifestiation d'intérêt: Projet *ArtSatel*," L'Observatoire de l'Espace du CNES. Available online: https://cnes-observatoire.fr/ode-projets/2024-02-19_artsatel/appel-a-projets-artsatel-2024.pdf (accessed March 21, 2024)

CNES (2024b), "La chute, appel à projets pour la création d'une œuvre en impesanteur," L'Observatoire de l'Espace du CNES. Available online: https://cnes-observatoire.fr/ode-projets/2024-03-06_appel-zero-g-chute/appel-a-projets-zero-g-chute.html (accessed March 21, 2024)

CNES (2024c), "Programme Avant-Poste La Chute, Appel à Projets de Création en Impesanteur," L'Observatoire de l'Espace du CNES. Available online: https://cnes-observatoire.fr/ode-projets/2024-03-06_appel-zero-g-chute/appel-a-projets-artistique-zero-g-chute.pdf accessed (accessed March 21, 2024)

Cohen, E. (2017), "The Paradoxes of Space Tourism," *Tourism Recreation Research*, 42 (1): 22–31.

Cole, N. (2020), "Boundary–Horizon–Landscape–Field (or The end of the World)," in P.J. Dunagan, M. Lazzara and N.J. Whittington (eds.), *Roots and Routes: Poetics at New College of California*, 53–7, Wilmington, Delaware: Vernon Press.

Coles, C. and A. Jefferson (1989), "The Space Chronometer: An Orbiting Clock for Humanity," *Leonardo*, 22 (2): 201–13.

Collett-White, M. (2009), "Banker-Turned-Artist Displays Zero Gravity Works," *Reuters*, January 13. Available online: https://www.reuters.com/article/us-banker-space/banker-turned-artist-displays-zero-gravity-works-idUSTRE50C42L20090113?sp=true (accessed August 7, 2018)

Colman, A. (2005), "Un/Becoming Digital: The Ontology of Technological Determinism and its Implications for Art Education," *The Journal of Social Theory in Art Education*, 25: 278–304.

Comsat (2019), 科幻电影《流浪地球》上映,与真实地球同框自拍. Available online: https://www.commsat.cn/news/read/101.html (accessed 16 October 2024)

Comte, P. (1987), "Leonardo in Orbit: Satellite Art," *Leonardo*, 20 (1): 17–21.

Coss, R.G., Y.A. Clearwater, C.G. Barbour and S.T. Towers (1989), "Functional Decor in the International Space Station: Body Orientation Cues and Picture Perception," National Aeronautics and Space Administration. Available online: https://ntrs.nasa.gov/api/citations/19900004615/downloads/19900004615.pdf (accessed March 20, 2024)

Crisp, F. and N. Triscot (2017), "Introduction," in F. Crisp and N. Triscott (eds.), *The Live Creature and Ethereal Things: Physics in Culture*, 9–11, London: Arts Catalyst.

D'Aloia, A. (2012), "The Intangible Ground: A Neurophenomenology of the Film Experience," *European Journal of Media Studies*, 1 (2): 219–39.

Dahl, E. (2010), "Towards a Phenomenology of Painting: Husserl's Horizon and Rothko's Abstraction," *Journal of the British Society for Phenomenology*, 41 (3): 229–45.

De La Cruz, P. (2014), "A Japanese Artist Launches Plants Into Space," *The New York Times*, July 18, 2014. Available online: https://archive.nytimes.com/tmagazine.blogs.nytimes.com/2014/07/18/flowers-in-space-azuma-makoto-exobiotanica/ (accessed December 11, 2023)

De Paulis, D. (2014). "AstroArts," *European Planetary Science Congress 2014*, EPSC Abstracts, 9: 839.

Deese, R.S. (2009), "The Artifact of Nature: 'Spaceship Earth' and the Dawn of Global Environmentalism," *Endeavour*, 33 (2): 70–5.

Dilworth, J. (2005), *The Double Content of Art*, New York: Prometheus Books.

Diner J. (2022), "NASA, Private Companies Count On Market Demand For Future Space Stations After ISS," *Space.com*. August 7. Available online: https://www.space.com/nasa-commercial-space-station-rely-on-market-demand (accessed March 17, 2024)

Donis, A.D. (1973), *A Primer of Visual Literacy*, Cambridge: MIT Press.

Du Preez, A. (2019), "Do Astronauts Dream of Post Earth?," *The Large Glass*, 27–8: 83–92.

Du Preez, A. (2022), *Art, the Sublime and Movement: Spaced Out*, New York: Routledge.

Dubois, K. (n.d.) "Creative Principles." Available online: https://www.kitsoudubois.com/english (accessed December 29, 2023)

Dubois, K. (1994), "'Dance and Weightlessness: Dancers' Training and Adaptation Problems in Microgravity,' *Leonardo*, 27 (1): 57–64.

Dubois, K. (2003 [1999]), "Body, Movement, Dance and Weightlessness," *Leonardo/Olats*, extracts from "The Application of Dance Techniques to

Weightless Flight: Weightless Dancer at the Art Aesthetics," PhD thesis. Sciences and Technologies department of the Université de Paris VIII. Available online: http://archive.olats.org/space/colloques/artgravitezero/t_Dubois_en.html (accessed December 29, 2023)

Dubois, K. (2005), "Kitsou Dubois," in N. Triscott and R. La Frenais (eds.), *Zero Gravity: A Cultural Users Guide*, 46–51, London: The Arts Catalyst.

Dunn, M. (2016), "A Year in Space: Aching Joints and the Weird Feeling of Clothing Touching Your Skin," *Global News*, March 4. Available online: https://globalnews.ca/news/2558552/ayear-in-space-aching-joints-and-the-weird-feeling-of-clothing-touching-your-skin/ (accessed March 22, 2024)

Dunnett, O., A.S. Maclaren, J. Klinger, K.M.D. Lane and D. Sage (2019), "Geographies of Outer Space: Progress and New Opportunities," *Progress in Human Geography*, 43 (2): 314–36.

During, E. (2019), "Evening Lecture—Philosophy in Zero G 2," European Graduate School Video Lectures, *YouTube*. Available online: https://www.youtube.com/watch?v=g_GaFggZI7s (accessed March 22, 2024)

Efland, A.D. (1988), "How Art Became a Discipline: Looking at Our Recent History," *Studies in Art Education*, 29 (3): 262–74.

Ekblaw, A. (2021), *Into the Anthropocosmos: A Whole Space Catalogue from the MIT Space Exploration Initiative*, Cambridge: MIT Press.

Entwistle, J. (2002), "The Dressed Body," in M. Evans and E. Lee (eds.), *Real Bodies*, 133–50, London: Palgrave.

Eshun, K. (2005), "On the Use and Abuse of Microgravity for Life," in N. Triscott and R. La Frenais (eds.), *Zero Gravity: A Cultural Users Guide*, 28–31, London: The Arts Catalyst.

Faust, J. (2022), "Orbital Reef Passes NASA Review," *Space News*, August 2. Available online: https://spacenews.com/orbital-reef-passes-nasa-review/ (accessed 3 January 2023)

Flusser, V. (2013 [1979]), *Natural: Mind*, trans R.M. Novaes, Minneapolis: Univocal.

Fong, J. (2014), "The Strange, Deadly Effects Mars Would Have on Your Body," *Wired*, February 11. Available online: https://www.wired.com/2014/02/happens-body-mars/ (accessed October 6, 2023)

Fortezza, R. (2001), "Facilities for Microgravity Fluid Science On Board the ISS,' in R. Monti (ed.), *Physics of Fluids in Microgravity*, 560–97, London: Taylor & Francis.

Frascina, F. (2000), "Introduction (1985)," in F. Frascina (ed.), 113–29, *Pollock and After: The Critical Debate*, 2nd edn., London: Routledge.

Frazer-Nash Consultancy (2021), "Space Based Solar Power: De-Risking the Pathway to Net Zero," Feasibility study for the UK Government Department for Business, Energy & Industrial Strategy. Available online: https://www.gov.uk/government/publications/space-based-solar-power-de-risking-the-pathway-to-net-zero (accessed January 8, 2023)

Freedman, L. (1997), "The 'Blurred' Horizon in Leonardo's Paintings," *Gazette Des Beaux-arts*, 129, May–June: 181–94.

Frost, R. (2012), "What Does it Sound Like Inside a Space Suit, While in Space?," *Quora*, November 23. Available online: http://www.quora.com/What-does-it-sound-like-inside-a-space-suit-while-in-space (accessed June 20, 2015)

Gallagher, S., L. Reinerman-Jones, B. Janz and J. Trempler (2015), *A Neurophenomenology of Awe and Wonder: Towards A Non-Reductionist Cognitive Science*, New York: Palgrave Macmillan.

Garriott, R. (2008), "Art in Space," *Our Space*. Available online: http://www.our-space.org/materials/states-of-matter/art-in-space (accessed January 23, 2017)

Gaskill, M. (2021), "Nine Ways We Use AR and VR on the International Space Station," National Aeronautics and Space Administration. Available online: https://www.nasa.gov/mission_pages/station/research/news/nine-ways-we-use-ar-vr-on-iss (accessed February 26, 2023)

Gast, M.A. and S.K. Moore (2011), "A Glimpse From the Inside of a Space Suit: What Is It Really like to Train for an EVA?," *Acta Astronautica*, 68 (1–2): 316–25.

Gérin, A. (2003), "Art in Space: Public Art in an Expanding World," *Espace Sculpture*, 64: 36–7.

Gever, E. (2016). "#LAUGH (2017)." Available online: http://www.eyalgever.com/laugh (accessed February 1, 2023)

Gov.uk (2022), "National Space Strategy." Available online: https://www.gov.uk/government/publications/national-space-strategy/national-space-strategy (accessed February 20, 2023)

Graham, S. (2016), *Vertical: The City from Satellites to Bunkers*, Brooklyn: Verso.

Graham, S. (2019), "Enigmatic Presence: Satellites and the Vertical Spatialities of Security," in Setha Low (ed.), *Spaces of Security: Ethnographies of Securityscapes, Surveillance, and Control*, 206–30, New York: New York University Press.

Griffin, J. (2018), "Orbital Reflector: The Artist Firing a Satellite into Space," *The Guardian*, November 26. Available online: https://www.theguardian.com/artanddesign/2018/nov/25/orbital-reflector-trevor-paglen-satellite (accessed January 18, 2023)

Groys, B. (2016), "Post-global Desire," in B. Groys (ed.), *8th Triennial of Contemporary Art U3: Beyond the Globe*, 12–15, Ljubljana: Sternberg Press.

Guarino, B. (2018), "Company Shoots Shiny Orb Into Orbit and Angers Astronomers Over Space Graffiti," *The Washington Post*, January 26. Available online: https://www.washingtonpost.com/news/speaking-of-science/wp/2018/01/26/company-shoots-shiny-orb-into-orbit-astronomers-irked-over-space-graffiti/ (accessed 18 January, 2023)

Guzman, C.R. (2018), "Mexican Technoscientific Arts, 2000–2015: Art and Science, Machine Inventions, and Political Ecologies." Masters thesis. City College of the City University of New York. Available online: https://academicworks.cuny.edu/cgi/viewcontent.cgi?article=1774&context=cc_etds_theses (accessed September 13, 2024)

Harrington, R. (2016), "A Famous Astronaut Says His Body Still Hurts after Spending A Year in Space," *Tech News*, June 8. Available online: https://www.businessinsider.com/scott-kelly-space-travel-effects-human-body-2016-6?amp (accessed March 8, 2024)

Heck, A. (2007), "Recent Ramblings in Astro-Art Relationship," *Memorie-Societa Astronomica Italiana*, 78 (1): 221.

Holden, K., M. Greene, E. Vincent, A. Sándor, A. Thompson, A. Feiveson and B. Munson (2022), "Effects of Long-duration Microgravity and Gravitational Transitions on Fine Motor Skills," *Human Factors*, 65 (6): 1046–58.

Holmes N.P., G.A. Calvert and C. Spence (2004), "Extending or Projecting Peripersonal Space with Tools? Multisensory Interactions Highlight Only the Distal and Proximal Ends of Tools," *Neurosci Lett*, 372 (1–2): 62–7.

Hooton, C. (2014), "A Bonsai Tree Has Been Launched Into Space," *The Independent*, July 22. Available online: https://www.independent.co.uk/news/world/americas/a-bonsai-tree-has-been-launched-into-space-9621630.html (accessed December 11, 2023)

Hoppenbrouwers, T., L. Ferra, M. Markus and M. Wolff (2017), "Operations Data Files—Driving Force Behind International Space Station Operations," *Acta Astronautica*, 138: 255–61.

Hughes, D. (1978), "Astro-art," *Nature*, 273: 326.

International Academy of Astronautics (2009), "Position Paper on The Role of Space Architecture."

Ingold, T. (2004), "Culture on the Ground," *Journal of Material Culture*, 9 (3): 315–40.

Ishiguro, S. (2004), "The 'Flying Deities Project': An Attempt at Performing East Asian Ancient Dances in Zero Gravity Environments," *Leonardo/Olats*. Available online: http://archive.olats.org/space/13avril/2004/te_sIshiguro.html (accessed December 28, 2023)

James, C. (2018), *Falling for Gravity: Invisible Forces in Contemporary Art*, Oxford: Peter Lang.

Jardetzky, W.C. (2005 [1958]), *Theories of Figures of Celestial Bodies*, Mineola: Dover.

JAXA (2014), "Koichi Wakata Weekly Activity Report, Vol. 9," *YouTube*. Available online: https://www.youtube.com/watch?v=AtqYemVvF7w (accessed January 19, 2024)

JAXA (2016), "Report on the Pilot Missions of Utilization for Culture/Humanities and Social Sciences (Summary Report)," JAXA-SP-15-011E. Available online: https://jaxa.repo.nii.ac.jp/record/3693/files/AA1530043000.pdf (accessed March 8, 2024)

Jelinek, A. (2020), *Between Discipline and a Hard Place: The Value of Contemporary Art*, London: Bloomsbury.

Johannsen, K. (2010), "Placelessness," *Nomadic Nature Kit: Book of Principles*. Available online: https://kirstenjohannsen.de/book-of-principles/principles/placelessness-example.html (accessed December 11, 2023)

Jörges, B. and J. López-Moliner (2017), "Gravity as a Strong Prior: Implications for Perception and Action," *Frontiers in Human Neuroscience*, April 28. Available online: https://www.ncbi.nlm.nih.gov/pmc/articles/PMC5408029/ (accessed March 29, 2024)

Kac, E. (1991), "Recent Experiments in Holopoetry and Computer Holopoetry," in T.H. Jeong (ed.), *Proceedings of the Fourth International Symposium on Display Holography (Proc. SPIE 1600)*, 229–36, Bellingham: SPIE. Available online: https://www.ekac.org/recent.experiments.html (accessed October 27, 2023)

Kac, E. (1997), "Key Concepts of Holopoetry," *Electronic Book Review*, January 3. Available online: https://electronicbookreview.com/essay/key-concepts-of-holopoetry/ (accessed February 17, 2023)

Kac, E. (2005), "Against Gravitropism: Art and the Joys of Levitation," in N. Triscott and R. La Frenais (eds.), *Zero Gravity: A Cultural Users Guide*, 18–25, London: The Arts Catalyst.

Kac, E. (2007), *Hodibis Potax*, Paris: Action Poétique.

Kac, E. (2014), "Inner Telescope: A Poem Written for and Realised In Zero Gravity." Available online: https://www.ekac.org/inner_telescope_english.html (accessed February 17, 2023)

Kac, E. (2021), *Télescope Intérieur: Les nouveau récits de l'espace*, Paris: Observatoire De L'Espace/CNES.

Karika, K. (2023), "Virgin Galactic," panel presentation at *Next-Generation Suborbital Researchers Conference*, Broomfield, Colorado, February 27–March 1.

Karmali, F. and M. Shelhamer (2008), "The Dynamics of Parabolic Flight: Flight Characteristics and Passenger Percepts," *Acta Astronautica*, 63 (5): 594–602.

Kleindienst, S. (2016), "Does the Cosmos Have a Center?," in B. Groys (ed.), K. Zakrajsek (trans.), *8th Triennial of Contemporary Art U3: Beyond the Globe*, 81, Ljubljana: Sternberg Press.

Kohl, K. (2020), "Memory Chain: A Pas de Deux of Artifact." Available online: https://www.katquay.com/pas-de-deux-of-artifact (accessed March 21, 2024)

Klinkrad, H. (2006), *Space Debris: Models and Risk Analysis*, Chichester: Praxis.

Kochetkova, E.S. (2020), "Horizon as a Symbolic Category in Contemporary Site-Specific Art," *Actual Problems of Theory and History of Art*, 10: 576–85.

Koek, A. (2017), "In/visible: The Inside Story of the Making of Arts at CERN," *Interdisciplinary Science Reviews*, 42 (4): 345–58.

Koenderink, J., A. van Doorn and K. Gegenfurtner (2017), "Compositorial 'Weight' & 'Luminance'," *Art & Perception*, 5 (3): 299–311.

Kotz, L. (2007), *Words to be Looked At: Language in 1960s Art*, Cambridge: MIT Press.

Kurbak, E. (2023), "Reinventing the Spindle: Politics of Craft in Space," *Proceedings of SIGGRAPH '23 Art Gallery*, New York: AMC.

La Frenais, R. (2005), "An Introduction to Vertigo," in N. Triscott and R. La Frenais (eds.), *Zero Gravity: A Cultural Users Guide*, 8–10, London: The Arts Catalyst.

Lau, C., C. Barriault and J. Krolik (2022), "Evaluating the Impact of a Comprehensive Canadian Science-Art Residency Program on the Participating Scientist, Artist and the Public," *Frontiers in Education*, 6. Available online: https://www.frontiersin.org/articles/10.3389/feduc.2021.690489/full (accessed March 20, 2024)

Lee, K.E. (2018), "Colonizing the Final Frontier: Why Space Exploration Beyond Low-Earth Orbit Is Central to US Foreign Policy, and the Legal Challenges It May Pose," *Southern California Interdisciplinary Law Journal*, 27 (1): 231–53.

Leslie, M. (2022), "Space Tourism Begins to Take Off," *Engineering*, 10 (March): 4–6.

Ling, A. (2020), "Work: Abiogenetic Triptych." Available online: https://andreasling.com/abiogenetic-triptych (accessed March 21, 2024)

Liu, X. (2020), "Sojourner 2020 | An international art payload to ISS," *MIT Media Lab*. Available online: https://www.media.mit.edu/posts/sojourner-2020/ (accessed March 10, 2023)

Liu, A. and P. Moran (2021), "Long Distance Low Gravity Love Letter: Telepresent Drawings in Space," in A. Ebklaw (ed.), *Into the Anthropocosmos*, 60–1, Cambridge: MIT Press.

Lockheed Martin (2024), "Nanoracks, Voyager Space, and Lockheed Martin Teaming To Develop Commercial Space Station." Available online: https://news.lockheedmartin.com/news-releases?item=129118 (accessed March 17, 2024)

Lopez-Alegria, M. (2023), "Ax-1," presentation at the Next-Generation Suborbital Researchers Conference, Broomfield, Colorado, February 27–March 1.

Maezawa, Y. (2021), "8 Crew Members Wanted: dearMoon—Special Message from MZ & Elon," *YouTube*. Available online: https://www.youtube.com/watch?v=h_L3wLmrMuA (accessed January 6, 2023)

Maezawa, Y. (2022), "A Message From MZ—dearMoon Crew Announcement," *YouTube*. Available online: https://www.youtube.com/watch?v=DKNSlL3Inn8 (accessed January 6, 2022)

Malina, R.F. (1991), "In Defense of Space Art: The Role of the Artist in Space Exploration," *International Astronomical Union Colloquium*, 112: 145–52.

Markov-Vetter, D. (2016), "Human Factors in Instructional Augmented Reality for Intravehicular Spaceflight Activities and How Gravity Influences the Setup of Interfaces Operated by Direct Object Selection," PhD Dissertation, University of Rostock, Germany.

Martínez, I. and Perales, J.M. (2001), "Mechanical Behavior of Liquid Bridges in Microgravity," in R. Monti (ed.), *Physics of Fluids in Microgravity*, 21–44, London: Taylor & Francis.

McCormick, T. (2021), "The Billionaire Space Race Reflects a Colonial Mindset That Fails To Imagine a Different World," *The Conversation*, August 15. Available online: https://theconversation.com/the-billionaire-space-race-reflects-a-colonial-mindset-that-fails-to-imagine-a-different-world-165235 (accessed 8 January, 2023)

Messeri, L. (2017), "We Need to Stop Talking About Space as a 'Frontier'," *Slate*, March 15. Available online: https://slate.com/technology/2017/03/why-we-need-to-stop-talking-about-space-as-a-frontier.html (accessed January 8, 2023)

Mir, A. (2017), *We Can't Stop Thinking about the Future: Artist Aleksandra Mir Speaks with the Space World*, London: Strange Attractor.

Monti, R. (2001), "Fluid Science in Microgravity Research," in Monti, R. (ed.), *Physics of Fluids in Microgravity*, 1–20, London: Taylor & Francis.

Murnik, M. (2016), "Art in the Environment of Zero Gravity: A Preliminary Sketch," *Metaverse Creativity*, 6 (1–2): 67–74.

Murnik, M. (2017), "Art in Weightlessness: From Outer Space to Virtual Worlds," *Virtual Creativity*, 7 (1): 29–40.

Myoda, P. (2013), "Artist Bradley Pitts Uses Space Travel As His Muse," *Vice*, December 3. Available online: https://www.vice.com/en/article/aenz98/artist-bradley-pitts-uses-space-travel-as-his-muse (accessed December 29, 2023)

Nahum (2015), "Holding Air," "Works," *Nahum*. Available online: https://nahum.xyz/works/holding-air/ (accessed September 13, 2024)

NASA (2013), "Acceleration Inside the International Space Station During a Reboost," *YouTube*. Available online: https://www.youtube.com/watch?v=5tcuKI3-B70 (accessed December 27, 2023)

NASA (2016), "Art Turns Public Eyes (and Ears) Toward Space," *NASA Global Climate Change News*. Available online: https://climate.nasa.gov/news/2507/art-turns-public-eyes-and-ears-toward-space/ (accessed December 11, 2023)

NASA (2021), "A Guide to Commercial Suborbital Flight Providers for Flight Opportunities," National Aeronautics and Space Administration. Available online: https://www.nasa.gov/sites/default/files/atoms/files/flight-providers-guide-2021-08.pdf (accessed January 6, 2023)

NASA (2022), International Space Station Transition Report. Available online: https://www.nasa.gov/wp-content/uploads/2015/01/2022_iss_transition_report-final_tagged.pdf (accessed November 5, 2022)

NASA (2023a), "The International Space Station Transition Plan," National Aeronautics and Space Administration. Available online: https://www.nasa.gov/faqs-the-international-space-station-transition-plan/#q1 (accessed March 17, 2024)

NASA (2023b), "International Space Station: Overview," National Aeronautics and Space Administration. Available online: https://www.nasa.gov/reference/international-space-station/ (accessed February 20, 2024)

NASA (2024), "Dark Matter & Dark Energy," National Aeronautics and Space Administration. Available online: https://science.nasa.gov/astrophysics/focus-areas/what-is-dark-energy (accessed March 17, 2024)

NASA Johnson (2012a), "Science off the Sphere: Space Soundwaves," *YouTube*. Available online: https://www.youtube.com/watch?v=U0rl_-z1YwQ (accessed December 24, 2023)

NASA Johnson (2012b), "Science off the Sphere: Astro Puffs," *YouTube*. Available online: https://www.youtube.com/watch?v=e6Faq1AmISI (accessed December 24, 2023)

NASA Johnson (2012c), "Science off the Sphere: Thin Film Physics," *YouTube*. Available online: https://www.youtube.com/watch?v=Uddz-3RwA_Y (accessed December 24, 2023)

NASA Johnson (2012d), "Science off the Sphere: Lenses and Vortices," *YouTube*. Available online: https://www.youtube.com/watch?v=CrTUpTDyGpI (accessed December 24, 2023)

NASA Johnson (2012e), "Science off the Sphere: Knitting Needle Experiment," *YouTube*. Available online: https://www.youtube.com/watch?v=qHrBhgwq__Q (accessed December 24, 2023)

NASA Johnson (2013), "NASA Astronaut Karen Nyberg Invites Quilters to Create a Space Square," *YouTube*, 31 October. Available online: https://www.youtube.com/watch?v=N0c1CijmH7o (accessed July 19, 2017)

Nelson, R. (2013), *Practice as Research in the Arts: Principles, Protocols, Pedagogies, Resistances*, New York: Palgrave Macmillan.

Niedderer, K. and L. Reilly (2010), "Research Practice in Art and Design: Experiential Knowledge and Organised Inquiry," *Journal of Research Practice*, 6 (2): Article E2.

Niedderer, K. and K. Townsend (2014), "Designing Craft Research: Joining Emotion and Knowledge," *Design Journal*, 17 (4): 624–48.

Noble, A. (2022), "*Rakia Mission 2022 Summary and Impact Report*." Available online: https://www.eng.rakiamission.co.il/blank-2 (accessed February 17, 2023)

Notarbartolo, A. (1975), "Some Proposals for Art Objects in Extraterrestrial Space," *Leonardo*, 8 (2): 139–41

Novarina, V. (2017), "Inner Telescope, a Space Artwork by Eduardo Kac (teaser)," *Vimeo*. Available online: https://vimeo.com/250786057 (accessed March 8, 2024)

Novikov, V.K., S.V. Nivikov and V.V. Tatarinov (2019), "Possible Directions for Reducing the Influence of the Rocket and Space Industry on the Environment," *AIP Conference Proceedings* 2171: 100003. Available online: https://aip.scitation.org/doi/epdf/10.1063/1.5133233 (accessed March 20, 2024)

O'Doherty, B. (1986), *Inside the White Cube: The Ideology of the Gallery Space*. Expanded edition. Berkeley: University of California Press.

O'Flaherty, K. and K. Isaak (2021), "Statement of Work," European Space Agency. Available online: https://www.cosmos.esa.int/documents/3867689/3867710/ARTSCI-EST-SOW-001-i1.0.pdf/8d3d2040-3752-61d3-195e-bc2cfc69bf23?t=1631791622116 (accessed March 20, 2024)

Ono, M. (2020), "Nothing, Something, Everything." Available online: https://cop21.net/ (accessed March 21, 2024)

Ormrod, J. and P. Dickens (2019), "Space Tourism, Capital, and Identity," *Space Tourism*, 25: 223–44.

Osaka, T. (2003), "The New Viewpoint from the World – Zero G," *Leonardo/Olats*. Available online: https://www.olats.org/takuro-osaka-the-new-viewpoint-from-the-world-zero-g/ (accessed February 25, 2024)

Osaka, T. (2004), "Art in the Space Age: Exploring the Relationship Between Outer Space and Earth Space," *Leonardo*, 37 (4): 273–6.

Osaka, T. (2007), "Marbling Painting on a Sphere of Water-I." Available online: https://takuro-osaka.com/en/space-art/space-art-marbling-painting-on/ (accessed December 26, 2023)

Osaka, T. (2011), "Marbling Painting–II 'Life Light and the Sea'." Available online: https://takuro-osaka.com/en/space-art/marbling-painting-on-a-sphere/ (accessed December 26, 2023)

Osaka, T. (2017), "Fusion of Science and Art 'Space Art-Light, Water, Life'," Special Lecture at University of Tsukuba. Available online: https://ocw.tsukuba.ac.jp/en/course/others/fusionofscienceandart2017/ (accessed October 8, 2023)

Paterson, C. (2015), "Matters of Gravity: An Interview with Mexican Space Artist Nahum," *The Nomadic Journal*, June 23. Available online: https://thenomadicjournal.com/2015/06/23/turning-art-upside-down-an-interview-with-mexican-space-artist-nahum/ (accessed 13 September 2024)

Pearlman, R. (2021), "That's a wrap! Soyuz lands with film crew after space station movie shoot," Space.com, October 17. Available online: https://www.space.com/soyuz-ms-18-landing-challenge (accessed December 28, 2021)

Peljhan, M. (2005), "Projekt Atol Flight Operations," in N. Triscott and R. La Frenais (eds.), *Zero Gravity: A Cultural Users Guide*, 32–7, London: The Arts Catalyst.

Pell, S.J., A.B. Imhof, C. Waldvogel, J.M. Kotler and M. Peljhan (2014), "Towards a Cooperation Between the Arts, Space Science Research and the European Space Agency—Preliminary Findings of the ESA Topical Team Arts and Sciences (ETTAS)," *Acta Astronautica*, 105: 181–91.

Perez, L. (2021), "Blue Origin, Virgin Galactic and the Kármán Line." Available online: https://aertecsolutions.com/en/2021/10/13/blue-origin-virgin-galactic-and-the-karman-line/ (accessed January 20, 2023)

Pettibone, L. (2018), "The Aesthetics of Gravity: How Our Relationship With Gravity is Revealed in Art," MA diss., Central Saint Martins College of Art and Design, University of the Arts, London.

Pietronigro, F. (2000), "Research Project Number 33: Investigating the Creative Process in a Microgravity Environment," *Leonardo*, 33 (3): 169–77.

Pietronigro, F. (2004), "Drift Painting in a Microgravity Environment and the Zero Gravity ArtsConsortium." *Leonardo/OLATS*. Available online: http://archive.olats.org/space/13avril/2004/te_fPietronigro.html (accessed January 20, 2023)

Pillai, S. (2024), "First Tests for Indian Space Station Expected Next Year: ISRO Chief," *Hindustan Times*, January 18. Available online: https://www.hindustantimes.com/india-news/first-tests-for-indian-space-station-expected-next-year-isro-chief-101705576078634.html (accessed March 17, 2024)

Pitts, B. (2011), "Singular Oscillations: One Step Removed," *YouTube*. Available online: https://www.youtube.com/watch?v=UO_-5idpJOU (accessed January 1, 2024)

Pocock, P. (2012), "Look Up! Art in the Age of Orbitization," in A.C.T. Geppert (ed.), *Imagining Outer Space: European Astroculture in the Twentieth Century*, 353–80, New York: Palgrave Macmillan.

Puncer, M. (2018), "Advanced Constructivism and Postgravity Art: Theoretical and Philosophical Implications," *Leonardo*, 51 (5): 475–81.

QAA (2019), "Subject Benchmark Statement: Art and Design." Available online: https://www.qaa.ac.uk/docs/qaa/subject-benchmark-statements/sbs-art-and-design-17.pdf?sfvrsn=71eef781_16 (accessed March 20, 2024)

Radford, T. (1999), "Hirst Painting Going to Mars," *The Guardian*, June 2. Available online: https://www.theguardian.com/uk/1999/jun/02/timradford1 (accessed March 18, 2022)

Raitt, D. (2007), "ESA and the Arts: A Programme in the Making," *Acta Astronautica*, 60: 48–54.

Ramiro, M. (2000), "Zero Gravity: A New Dimension for the Object, in A. Kostiæ (ed.), *I Levitate, What's Next …*, 98–112, Maribor, Slovenia: Association for Culture and Education KIBLA. Available online: http://multimedia.kibla.org/doc/knjiga_levitacija.pdf (accessed March 21, 2024)

Redzisz, K. (2015), "Maria Bartuszová, Untitled (Drop), 1963–4," *Tate Modern*. Available online: https://www.tate.org.uk/art/artworks/bartuszova-untitled-drop-t14516 (accessed April 5, 2023)

Richmond, S. (2015), *Lynda Benglis: Beyond Process*, London: Bloomsbury.

Robinson, J. (2009), "From Abstraction to Model: George Brecht's Events and the Conceptual Turn in Art of the 1960s," *October*, 127: 77–108.

Roca, M.A. (2005), "Dédalus Project Microperformances, 2003," in N. Triscott and R. La Frenais (eds.), *Zero Gravity: A Cultural Users Guide*, 74–7, London: The Arts Catalyst.

Rogers, M., G. Vogt and M. Wargo (2009), "Microgravity: A Teacher's Guide With Activities in Science, Mathematics, and Technology," National Aeronautics and Space Administration. Available online: https://www.nasa.gov/wp-content/uploads/2009/07/62474main_microgravity_teachers_guide.pdf (accessed December 26, 2023)

Rudolf, A. (1974), *Art and Visual Perception*, Berkeley: University of California Press.

Ryan, R.G., E.A. Marais, C.J. Balhatchet and S.D. Eastham (2022), "Impact of Rocket Launch and Space Debris Air Pollutant Emissions On Stratospheric Ozone and Global Climate," *Earth's Future*, 10 (6): e2021EF002612.

Rycroft, S. (2019), "The Artist Placement Group: An Archaeology of Impact," *Cultural Geographies*, 26 (3): 289–304.

Ryklin, M. (2005), "Being There," in N. Triscott and R. La Frenais (eds.), *Zero Gravity: A Cultural Users Guide*, 14–17, London: The Arts Catalyst.

Sagan, C. (1994), *Pale Blue Dot*, New York: Ballantine.

Sato, S. (2014), "Cherry Tree From Space Mystery Baffles Japan," *Phys.org*, April 11. Available online: https://phys.org/news/2014-04-cherry-tree-space-mystery-baffles.html (accessed September 16, 2023)

Savage, A. (2015), "Adam Savage Incognito at Comic-Con 2015 (with Astronaut Chris Hadfield)," *Tested*. Available online: http://www.tested.com/art/makers/534134-adam-savage-incognito-comic-con-2015-astronaut-chris-hadfield/ (accessed 18 July, 2015)

Schrader, A. (2023), "A Global Project Is Sending Time Capsules to the Moon Containing Artworks and Objects by More Than 30,000 Creatives," *Artnet*, July 31. Available online: https://news.artnet.com/art-world/lunar-codex-collection-art-poetry-will-be-sent-to-moon-2343877#:~:text=It%20featured%20drawings%20by%20artists,Brunetti%2C%20Pearl%20(2020) (accessed March 25, 2024)

Schwartz, A.E. (2009), "Rising above the Horizon: Visual and Conceptual Modulation of Space and Place," *Augenblick*, 45: 36–61.

Schwartz, A.E. (2010), "Rising above the Horizon: Visual and Conceptual Modulation of Space and Place," *AugenBlick. Marburger Hefte zur Medienwissenschaft*, 45: 36–61.

Segal, L. and Y. Meroz (2022), "Impossible Object: Art in Space." Available online: https://www.liatsegal.com/ (accessed January 19, 2023)

Sethi, V. (2022), "'The Painter's New Tools' Reimagines the Traditional Medium of Painting," *STIRworld*, September 5. Available online: https://www.stirworld.com/see-news-the-painters-new-tools-reimagines-the-traditional-medium-of-painting (accessed February 25, 2024)

Sharma, S. (2020), "Fluid Expressions in Microgravity," *Space Exploration*, MIT Media Lab. Available online: https://zero-gravity.pubpub.org/pub/fluid-expressions/release/6 (accessed December 24, 2023)

Sharma, S. (2022), "Zero G Lightening Talks," MIT Space Exploration Initiative Beyond the Cradle 2022, April 28, *YouTube*. Available online: https://www.youtube.com/watch?v=FiXDjqSVHdU&t=4543s (accessed January 12, 2024)

Sharma, S. (2023), "Fluid Expressions." Available online: https://www.sanasharma.com/fluid-expressions (accessed December 24, 2023)

Shatner, W. and J. Brandon (2022), *Boldly Go: Reflections on a Life of Awe and Wonder*, New York: Atria Books.

Shepherd, T. (2023), "Picture Imperfect: Light Pollution from Satellites is Becoming an Existential Threat to Astronomy," *The Guardian*, January 5. Available online: https://www.theguardian.com/science/2023/jan/06/picture-imperfect-light-pollution-from-satellites-is-becoming-an-existential-threat-to-astronomy (accessed 18 January 2023)

Sherwood, B. (2010), "Module Architecture for In Situ Space Laboratories," *AIAA Meeting Paper: 40th International Conference on Environmental Systems*,

July 11–15, 2010. Available online: https://doi.org/10.2514/6.2010-6048 (accessed March 29, 2024)

Sim, T. (2020), "The Cinematography of The Expanse | Camera & Lighting Breakdown," Indy Mogul, *YouTube*. Available online: https://www.youtube.com/watch?v=29MzR2m88G8 (accessed October 18, 2023)

Singapore Space (2017), "2015 Try Zero G Experiment—Zero G painting," *YouTube*. Available online: https://www.youtube.com/watch?v=cgawkgdL5Ms (accessed January 19, 2024)

Smith, L. (2022), "Augmented Reality in Microgravity," paper presented at *AWE USA 2022, XR Enablement. YouTube*. Available online: https://www.youtube.com/watch?v=-KazzqZHNM8 (accessed February 26, 2023)

Smith, W. (2009), "Conclusion: To Infinity and Beyond?," in D. Bell and P. Martin (eds.), *Space Travel & Culture: From Apollo to Space Tourism*, 204–12, Oxford: Blackwell.

Sobchack, V. (1982), "Toward Inhabited Space: The Semiotic Structure of Camera Movement in the Cinema," *Semiotica*, 41 (1–4): 317–35.

Spaceskin (2022), "SpaceTouch," *MIT Media Lab*. Available online: https://www.media.mit.edu/projects/spaceTouch/overview/ (accessed January 1, 2024)

Spencer, J. (2022), *Statement at the Overview Roundtable, December 21, 2022*. Overview Round Table, The Human Space Program.

Spennemann, D.H.R. (2008), "Orbital Lunar and Interplanetary Tourism," in C. Marin and J. Jafari (eds.), *Starlight: A Common H eritage*, 363–75, San Cristóbal de La Laguna: Instituto de Astrofisica de Canárias.

Steyerl, H. (2011), "In Free Fall: A Thought Experiment on Vertical Perspective," *E-flux*, 24. Available online: https://www.e-flux.com/journal/24/67860/in-free-fall-a-thought-experiment-on-vertical-perspective/ (accessed August 19, 2022)

Stott, N. (2019), "What it's Like to Paint in Space—According to a NASA Astronaut," *Quartz*, March 23. Available online: https://qz.com/1578231/what-its-like-to-paint-in-space-according-to-a-nasa-astronaut (accessed December 24, 2023)

Struyanskiy, A. (2022), "Submission: Human Frontiers," *Moon Gallery*. Available online: https://moongallery.eu/artist/arti-struyanskiy/ (accessed February 26, 2023)

Sumini, V. (2021), "SpaceHuman 2.0," *MIT MediaLab*. Available online: https://www.media.mit.edu/projects/spacehuman/overview/ (accessed January 1, 2024)

Sutton, G. (2019), "Hans Haacke: Works of Art, 1963–72," in M. Gioni and G. Carrion-Murayari (eds.), *Hans Haacke: All Connected*, 10–15, New York: Phiadon.

Taibu, R., D. Rudge and D. Schuster (2015), "Textbook Presentations of Weight: Conceptual Difficulties and Language Ambiguities," *Physical Review Special Topics: Physics Education Research*, 11 (1): 0101171–01011720.

Tate Modern (2022), "Maria Bartuszová: Every Tangle of Thread and Rope," exhibition booklet.

Taylor, R. (2003), "Rebecca Horn: Concert for Anarchy," *Tate online*. Available online: https://www.tate.org.uk/art/artworks/horn-concert-for-anarchy-t07517 (accessed February 16, 2023)

The Arts Catalyst (2005), Executive Summary: Cultural Utilisation of the International Space Station. Available online: https://i-dat.org/wp-content/blogs.dir/28/files/2018/11/ISSExecSummary.pdf (accessed February 22, 2023)

Thornton, W.E. (1978), "Anthropometric Changes in Weightlessness," in E. Churchill, L.L. Laubach, J.T. Mcconville and I. Tebbets (eds.), *Anthropometric Source Book: Volume 1: Anthropometry for Designers*, i1–i105, National Aeronautics and Space Administration. Available online: https://ntrs.nasa.gov/api/citations/19790003563/downloads/19790003563.pdf (accessed March 10, 2024)

Today (n.d.), "William Shatner Reacts To Seeing Earth From Space: 'It's So Fragile'," *YouTube*. Available online: https://www.youtube.com/watch?v=bx_CdBcRexc (accessed December 13, 2025)

Todorov, T. (2014), *Elemental Sculpture: Theory and Practice*, Newcastle: Cambridge Scholars Publisher.

Tolfo, G., E. Griffin, M. Ridge, K. Beelen and R. Ahnert (2024), "The Minimum Research Outcomes: A Mechanism for Generating and Managing Projects in Labs," in U. Pawlicka-Deger and C. Thompson (eds.), *Digital Humanities and Laboratories: Perspectives on Knowledge, Infrastructure and Culture*, New York: Routledge.

Tosa, N., A. Yamada, Y. Pang, S. Toba, A. Ito, T. Suzuki and R. Nakatsu (2023), "Sound of Ikebana: Fluid Artwork Created under Zero-G Using Parabolic Flight," *Leonardo*, 56 (4): 359–66.

Triscott, N. (2012), "Spaced Out … The Relationship Between Art and Space Agencies." Available online: https://nicolatriscott.org/2012/03/23/spaced-out-the-relationship-between-art-and-space-agencies/ (accessed February 26, 2023)

Triscott, N. (2015), "High Energy Interactions: How to be an Artist at CERN." Available online: https://nicolatriscott.org/2015/06/27/high-energy-interactions-how-to-be-an-artist-at-cern/ (accessed March 1, 2023)

Triscott, N. and R. La Frenais, eds. (2005), *Zero Gravity: A Cultural Users Guide*, London: The Arts Catalyst.

Tsong-Zung, C. and G. Shiming (2016), "Yaji Garden: Art Under the Sky," in B. Groys (ed.), K. Zakrajsek (trans.), *8th Triennial of Contemporary Art U3: Beyond the Globe*, 132–41, Ljubljana: Sternberg Press.

Turšič, M. (2013), "Space Culturalization," TEDxLjubljana, *YouTube*. Available online: https://www.youtube.com/watch?v=XF2McMNvbkI (accessed January 17, 2023)

Ulrich, B. (2007), "NASA and the Arts," *50th Magazine—50 Years of Exploration and Discovery*, National Aeronautics and Space Administration. Available online: https://www.nasa.gov/50th/50th_magazine/arts.html (accessed February 26, 2023)

US Department of Commerce (1932), "The International Metric System of Weights and Measures. Miscellaneous Publication no. 135." Available online: https://www.govinfo.gov/content/pkg/GOVPUB-C13-7bce1f6babbcbb82fe740fbde6ff1738/pdf/GOVPUB-C13-7bce1f6babbcbb82fe740fbde6ff1738.pdf (accessed October 13, 2023)

Usborne, S. (2009), "Close Up: Nasser Azam," *The Independent on Sunday*, January 4. Available online: https://www.independent.co.uk/arts-entertainment/art/features/closeup-nasser-azam-1219710.html (accessed March 10, 2024)

vanEenoo, C. (2012), "The Art of Making Mistakes," *International Journal of Arts*, 2 (3): 7–10.

van Linden Tol, A. (2017), "The Collision of Art & Science," presentation at Deep Space 8K, *Ars Electronica*, June 22. Available online: https://www.youtube.com/watch?v=mXdsNLiOLEw (accessed February 26, 2023)

Verhoeven, A. (2011), "Craft as an Attitude," *Studio Magazine*, Fall–Winter 2011–12: 12–16.

Watlington, E. (2023), "The Best and Worst Artists of Our Time are Sending Their Work into Space," *Art in America*, May 16. Available online: https://www.artnews.com/art-in-america/columns/space-art-jeff-koons-xu-bing-xin-liu-1234667481/ (accessed September 14, 2023)

Webber, D. (2019), "Current space Tourism Development," in E. Cohen and S. Spector (eds.), *Space Tourism: The Elusive Dream*, 163–76, Bingley: Emerald.

Weddigen, T. (2018), "Alien Spotting: Damien Hirst's Beagle 2 Mars Lander Calibration Target and the Exploitation of Outer Space," in Geppert, A. (ed.), *Imagining Outer Space*. Palgrave Studies in the History of Science and Technology, 335–49, Palgrave Macmillan, London.

White, F. (2021), *The Overview Effect: Space Exploration and Human Evolution*. 4th edn, Denver: Multiverse Publishing.

White, F. (2022), "The Overview Effect and Suborbital Flights," paper presented at *2023 Next-Generation Suborbital Researchers Conferenc*e, Broomfield, Colorado, February 27 to March 1, 2023.

Wightman, M. (2005), "Falling without Fear," in N. Triscott and R. La Frenais (eds.), *Zero Gravity: A Cultural Users Guide*, 62–65, London: The Arts Catalyst.

Wilson, S. (2002), *Information Arts: Intersections of Art, Science, and Technology*, Cambridge: MIT Press.

Woods, A.R. (1993), "*The Cosmic Dancer*: Sculpture and the Absence of Gravity," *Leonardo*, 26 (4): 297–301.

Woods, A. (2005), "The Cosmic Dancer: Sculpture and the Absence of Gravity," *The OURS Project*. Available online: http://www.ours.ch/cosmicdancer.php (accessed February 26, 2023)

Woods, A. (2019), "Cosmic Dancer Sculpture to the Mir Space Station," *The Space Option*, October 6. Available online: https://thespaceoption.com/the-spaceflight-of-the-cosmic-dancer-sculpture-to-the-mir-space-station/ (accessed March 10, 2024)

Woods, A. (2021a), "Cosmic Dancer," video for the *2021 Space Renaissance Congress Art Gallery*: *Session: 5.2 Space Renaissance Art*. Available online: https://www.cosmicdancer.com/cosmic_dancer_project_videos.php (accessed October 31, 2022)

Woods, A. (2021b), "Arthur R. Woods—AstroArtist—TerraArtist," *Cosmic Dancer*. Available online: https://www.cosmicdancer.com/arthur_woods_artist_biographical_information.php (accessed January 17, 2023)

Woods, A. (2022), "Art to the Stars: An Astronautical Perspective on the Arts and Space," *Greater Earth*. Available online: https://greater.earth/ART_DOCS/art_to_the_stars.php (accessed March 10, 2024).

Yang, T., J.R. Kim, H. Jin, H. Gil, J. Koo and H.J. Kim (2021), "Recent Advances and Opportunities of Active Materials for Haptic Technologies in Virtual and Augmented Reality," *Advanced Functional Materials*, 31 (39). Available online: https://onlinelibrary.wiley.com/doi/full/10.1002/adfm.202008831 (accessed February 27, 2024)

Zaklynsky, A. (2018), "Developing Structures for an International Art Gallery on the Moon," paper presented at *42nd COSPAR Scientific Assembly*, July 14–22, Pasadena, California, USA. Available online: https://zaklynsky.com/index.php?/press–article/42nd-cospar-scientific-assembly/ (accessed October 16, 2023)

Živadinov, D. (2010), 50 Coordinates of Postgravity Art. Available online: https://www.scribd.com/document/31097592/50-kordinat (accessed December 12, 2024)

Žiberna, M. (2011), "Theatre Director Enraptured by Space," *The Slovenia Times*, December: 38–9.

INDEX